DIGITAL TECHNOLOGY AND THE PRACTICES OF HUMANITIES RESEARCH

Digital Technology and the Practices of Humanities Research

Edited by Jennifer Edmond

https://www.openbookpublishers.com

© 2020 Jennifer Edmond. Copyright of individual chapters is maintained by the chapters' authors.

This work is licensed under a Creative Commons Attribution 4.0 International license (CC BY 4.0). This license allows you to share, copy, distribute and transmit the work; to adapt the work and to make commercial use of the work providing attribution is made to the author (but not in any way that suggests that they endorse you or your use of the work).

Attribution should include the following information:

Jennifer Edmond (ed.), *Digital Technology and the Practices of Humanities Research*. Cambridge, UK: Open Book Publishers, 2020, https://doi.org/10.11647/OBP.0192

In order to access detailed and updated information on the license, please visit https://doi.org/10.11647/OBP.0192#copyright

Further details about CC BY licenses are available at http://creativecommons.org/licenses/by/4.0/

All external links were active at the time of publication unless otherwise stated and have been archived via the Internet Archive Wayback Machine at https://archive.org/web

Any digital material and resources associated with this volume are available at https://doi.org/10.11647/OBP.0192#resources

Every effort has been made to identify and contact copyright holders and any omission or error will be corrected if notification is made to the publisher.

ISBN Paperback: 978-1-78374-839-6
ISBN Hardback: 978-1-78374-840-2
ISBN Digital (PDF): 978-1-78374-841-9
ISBN Digital ebook (epub): 978-1-78374-842-6
ISBN Digital ebook (mobi): 978-1-78374-843-3
ISBN Digital (XML): 978-1-78374-844-0
DOI: 10.11647/OBP.0192

Cover image: photo by Nanda Green on Unsplash https://unsplash.com/photos/BeVWHMXYwwo
Cover design: Anna Gatti

Contents

Acknowledgements		ix
Notes on the Contributors		xi
1.	**Introduction: Power, Practices, and the Gatekeepers of Humanistic Research in the Digital Age**	1
	Jennifer Edmond	
	The Impact of Collaboration	13
	Evaluators as Gatekeepers	14
	Publishers as Gatekeepers	16
	This Volume's Contribution	18
	Bibliography	19
2.	**Publishing in the Digital Humanities: The Treacle of the Academic Tradition**	21
	Adriaan van der Weel and Fleur Praal	
	The Functions of Scholarly Publishing in the Print Paradigm	25
	Transferring the Functions of Publishing to the Digital Medium	29
	Dissemination	31
	Registration	34
	Certification	38
	Archiving	40
	Conclusions	41
	Bibliography	44
3.	**Academic Publishing: New Opportunities for the Culture of Supply and the Nature of Demand**	49
	Jennifer Edmond and Laurent Romary	
	Introduction	49
	The Place of the Book in Humanities Communication	52
	Scholarly Reading and Browsing	55

	Old and New Ways to Share Knowledge	58
	The Evaluator as an Audience for Scholarship	62
	Barriers to Change, and Opportunities	63
	Research Data and the Evolving Communications Landscape	71
	Conclusions	72
	Bibliography	75
4.	**The Impact of Digital Resources**	**81**
	Claire Warwick and Claire Bailey-Ross	
	Understanding and Measuring Impact	82
	Commercial Impact	91
	Media and Performance	92
	Cultural Heritage	93
	Policy Impact	96
	Limitations of the REF Case Studies	96
	Conclusions	98
	Bibliography	99
5.	**Violins in the Subway: Scarcity Correlations, Evaluative Cultures, and Disciplinary Authority in the Digital Humanities**	**105**
	Martin Paul Eve	
	Judging Excellence and Academic Hiring and Tenure	107
	The Diverse Media Ecology of Digital Humanities	112
	Strategies for Changing Cultures: Disciplinary Segregation, Print Simulation, and Direct Economics	115
	Bibliography	119
6.	**'Black Boxes' and True Colour — A Rhetoric of Scholarly Code**	**123**
	Joris J. van Zundert, Smiljana Antonijević, and Tara L. Andrews	
	Introduction	123
	Background	125
	Methodology	131
	Experiences	134
	Inventio — The Impetus for DH Researchers to Code	134
	Dispositio — How Coding Constructs Argument	137
	Elocutio — Coding Style, Aesthetics of Code	141
	Memoria — The Interaction between Code and Theory	143
	Actio — The Presentation and Reception of DH Codework	146

	Conclusions	150
	Recommendations	152
	Appendix 6.A: Survey Questions	157
	Bibliography	158
7.	**The Evaluation and Peer Review of Digital Scholarship in the Humanities: Experiences, Discussions, and Histories**	**163**
	Julianne Nyhan	
	Introduction	163
	Experiences and Discussion of Evaluation c. 1963–2001	167
	Individual and Group Experiences of Making Digital Scholarship	168
	What Should Be Evaluated?	170
	Which Evaluative Criteria?	172
	Organising the Peer Review Process	173
	Implicit Peer Review	174
	Conclusion	177
	Bibliography	179
8.	**Critical Mass: The Listserv and the Early Online Community as a Case Study in the Unanticipated Consequences of Innovation in Scholarly Communication**	**183**
	Daniel Paul O'Donnell	
	The Listserv as Case Study	185
	You've got Mail	186
	The LISTSERV Revolution	188
	The Invisible Seminar	189
	The Invisible Water-Cooler	191
	What Is It that an Academic Mailing List Disrupts?	195
	Online Communities vs Learned Societies	198
	Same as it Ever Was? Looking Backwards and Forwards	200
	Conclusion	202
	Bibliography	203
9.	**Springing the Floor for a Different Kind of Dance: Building DARIAH as a Twenty-First-Century Research Infrastructure for the Arts and Humanities**	**207**
	Jennifer Edmond, Frank Fischer, Laurent Romary, and Toma Tasovac	
	Introduction: What's in a Word?	207
	But What *Is* Research Infrastructure?	210

Infrastructures as Knowledge Spaces	212
Why Do the Arts and Humanities Need Research Infrastructure?	214
History of a New Model of RI Development	216
The Activities of the DARIAH ERIC	221
The DARIAH Marketplace	222
DARIAH Working Groups	225
Policy and Foresight	225
Training, Education, Skills, and Careers	226
Conclusions (and a Few Concerns)	227
Appendix 9.A: Definitions of Research Infrastructure	230
Bibliography	232

10. **The Risk of Losing the *Thick Description*: Data Management Challenges Faced by the Arts and Humanities in the Evolving FAIR Data Ecosystem** — 235

    Erzsébet Tóth-Czifra

Realising the Promises of FAIR within Discipline-Specific Scholarly Practices	235
A Cultural Knowledge Iceberg, Submerged in an Analogue World	237
Legal Problems that Are Not Solely Legal Problems	239
The Risk of Losing the Thick Description upon the Remediation of Cultural Heritage	242
The Scholarly Data Continuum	247
Data in Arts and Humanities — Still a Dirty Word?	250
The Critical Mass Challenge and the Social Life of Data	251
The Risk of Losing the *Thick Description* — Again	255
Conclusions: On our Way towards a Truly FAIR Ecosystem for the Arts and Humanities	258
Bibliography	263

Index — 267

Acknowledgements

First and foremost, the editor of this volume would like to thank the European Science Foundation for making possible both the original working group along with its meetings, and this open access publication. The NeDiMAH network continues to be a point of reference for scholars who are exploring not just how to use digital methods in the humanities and what it means to do this, but also what is at stake in the digital turn for our diverse and yet interconnected disciplines.

It would be remiss not to also thank the participants in the NeDiMAH events: their contributions to that early discussion are woven into the fabric of this volume and the issues it pursues. In particular, I would like to thank the Zadar meeting group: Linda Bree, Emma Clarke, Marin Dacos, Bianca Gualandi, Angela Holzer, Christina Kamposiori, Eva Kekou, Camilla Leathem, Francesca Morselli, Claudine Moulin, Alex O'Connor, Franjo Pehar, and Susan Reilly. Their collective and enthusiastic commitment to capturing a multidisciplinary and multisectoral snapshot of the shifts occurring in the communications landscape of the arts and humanities remain astonishingly relevant even after so many years. Finally, I am grateful to the many authors of this work who have either been required to show great patience with the slow development of the volume or work to very tight deadlines in order to bring its slow-growing content up to date. I include in this group those authors who were, for a number of reasons, unable to stay with the volume until the end, but whose drafts contributed to my own understanding of the institutional and individual issues in play. In particular, I would like to warmly thank Susan Schreibman for her early contributions in clarifying the focus of this volume and assembling an exciting panel of contributors and Laurence Taylor for his careful copyediting.

Notes on the Contributors

Tara L. Andrews is a university professor in Digital Humanities at the University of Vienna. She obtained her DPhil in Oriental Studies at the University of Oxford in 2009; she also holds a MPhil in Byzantine Studies from Oxford, and a BSc in Humanities and Engineering from the Massachusetts Institute of Technology. She currently leads an SNSF-funded project to produce a digital critical edition of the twelfth-century Armenian-language chronicle by Mattēos Uṙhayeci (Matthew of Edessa). More broadly, Andrews's research interests include Byzantine history of the middle period (in particular the tenth to twelfth centuries), Armenian history and historiography from the fifth to the twelfth centuries, and the application of computational analysis and digital methods to the fields of medieval history and philology. From 2010 to 2013, Andrews worked at the KU Leuven University with Prof. Caroline Macé on the 'Tree of Texts' project, which is an investigation of the theory behind the stemmatic analysis of classical and medieval manuscript texts. The suite of online tools developed for the project is freely available online at https://stemmaweb.net/stemmaweb/

Smiljana Antonijević explores the intersection of communication, culture, and technology through research and teaching in the USA and Europe. She is the author of *Amongst Digital Humanists: An Ethnographic Study of Digital Knowledge Production* (2015), while other recent publications include *Developing Tools for Voices from the Field* (2016), *Personal Library Curation* (2014), *Working in Virtual Knowledge* (2013), *The Immersive Hand* (2013), *Cultures of Formalization* (2012), and *Researchers' Information Uses in a Digital World* (2012). Antonijević's most recent research projects are Alfalab: eHumanities Tools and Resources, Humanities Information Practices, Digitizing Words of Power, and Digital Scholarly Workflow.

Claire Bailey-Ross is a senior lecturer in user experience at the School of Creative Technologies, University of Portsmouth. She is course leader for the BSc Digital Media Degree Programme. Bailey-Ross' research takes place within the context of digital humanities and her work is highly interdisciplinary: ranging from the user's experience of digital heritage resources to broader debates surrounding the impact of digital innovation and technological change in cultural heritage environments. Her current research interests include the nature of participation and engagement provided by digital technology, knowledge transfer between academic and cultural heritage institutions, and the innovation opportunities afforded by humanities research.

Jennifer Edmond is an associate professor of Trinity College Dublin and a co-director of the Trinity Centre for Digital Humanities. She holds a PhD in Germanic Languages and Literatures from Yale University, and applies her training as a scholar of language, narrative, and culture to the study and promotion of advanced methods in, and infrastructures for, the arts and humanities. In this vein, Edmond serves as president of the board of directors of the pan-European research infrastructure for the arts and humanities DARIAH-EU. Additionally, she represents this body on the Open Science Policy Platform (OSPP), which supports the European Commission in developing and promoting Open Science policies. She has also developed a significant individual profile within European research and research policy circles in the past five years, having been named one of Ireland's five 'Champions of EU Research' in 2012. She coordinated the €6.5m CENDARI FP7 (2012–2016) project and is a partner in the related infrastructure cluster PARTHENOS. Edmond was also coordinator of the 2017–2018 ICT programme-funded project KPLEX, which investigated bias in big data research from a humanities perspective, and she is currently a partner in the CHIST-ERA project PROVIDEDH, which is investigating progressive visualisation as support for managing uncertainty in humanities research.

Martin Paul Eve is the Chair of Literature, Technology, and Publishing at Birkbeck, University of London. He is the author of four books, including *Open Access and the Humanities: Contexts, Controversies and the Future* (2014) and is one of the founders of the Open Library of Humanities.

Frank Fischer is currently an associate professor in digital humanities at the Higher School of Economics, Moscow, and co-director of DARIAH-EU. He has studied computer science, German literature, and Spanish philology in Leipzig and London, and is an *Ancien Pensionnaire de l'École Normale Supérieure* in Paris. He received his PhD from the University of Jena for his study on the dramatic works of Joachim Wilhelm von Brawe and their contemporary translations into Russian, Danish, and French.

Julianne Nyhan is an associate professor in Digital Information Studies of the Department of Information Studies, UCL, where she leads the Digital Humanities MA/MSc Programme. Nyhan is also the deputy director of the UCL Centre for Digital Humanities and on the leadership group of the UCL Centre for Critical Heritage. She has published widely on the history of digital humanities and her work has been translated into a number of languages, including Russian, Polish, and Chinese. Recent publications include (with Andrew Flinn) *Computation and the Humanities: Towards an Oral History of Digital Humanities* (2016). Her research projects include a Leverhulme-funded collaboration with the British Museum on the manuscript catalogues of Sir Hans Sloane; an ESRC-funded historical newspaper data mining project; and a Marie Skłodowska-Curie action called 'Critical Heritage Studies and the Future of Europe'.

Daniel Paul O'Donnell is a professor of English and an adjunct member of the Library Research Faculty at the University of Lethbridge. His research and teaching interests include the digital humanities, scholarly and scientific communication, textual and editorial theory and practice, globalisation, and Anglo-Saxon studies. He is the Editor in Chief of the open access journal *Digital Studies/Le Champ Numérique*, and president of Force11.org. In previous years he was the founding chair of both the Digital Medievalist and Global Outlook::Digital Humanities as well as being president of the Text Encoding Initiative. His digital edition of the Anglo-Saxon poem *Cædmon's Hymn* received an honourable mention in the MLA's prize for most distinguished scholarly edition of 2006.

Fleur Praal graduated in 2012 with a MA cum laude in Book and Digital Media Studies at Leiden University with a thesis on the quantitative analysis of libraries and publishers' data in the Netherlands. After gaining experience in quantitative research evaluation at Leiden's Centre for Science and Technology Studies, she returned to the Book and Digital Media Studies department as a PhD researcher and lecturer specialising in publishing studies. Her dissertation combines methodologies and models from book and publishing studies, cultural analysis, and the sociology in analysing the changing landscape of scholarly publishing in the current digital age, and especially in the humanities.

Laurent Romary is *directeur de recherche* at Inria (France), within the ALMAnaCH team, and a former director general of DARIAH (2014–2018). He carries out research on the modelling of semi-structured documents with a specific emphasis on texts and linguistic resources. He has been active in standardisation activities within the ISO committee TC 37 and the Text Encoding Initiative. Romary has also been working for many years on the advancement of open access.

Toma Tasovac is director of the Belgrade Center for Digital Humanities (BCDH) and a member of the DARIAH-EU board of directors. His areas of interest include lexicography, data modelling, the TEI (Text Encoding Initiative), digital editions, and research infrastructures. Tasovac serves on the advisory board for Europeana Research, and on the CLARIN-DE/DARIAH-DE technical board. He is also a steering group member for the European Network of eLexicography (ENeL) and the European Network for Combining Language Learning with Crowdsourcing Techniques (enetCollect). Under Tasovac's leadership, BCDH has received funding from various national and international grant bodies, including Erasmus Plus and Horizon 2020.

Erzsébet Tóth-Czifra received her PhD in cultural linguistics in 2018 at Eötvös Loránd University. In 2016, her commitment to democracy in science led her to join the research discovery platform ScienceOpen, and begin her carrier as an open science advocate. Currently she works as open science officer at the European Research Infrastructure Consortium DARIAH where she contributes to the design and implementation of open science policy statements, guidelines, and services related to the open dissemination of research results in the humanities.

Adriaan van der Weel is Bohn Extraordinary Professor of Book Studies, and teaches book and digital media studies at the University of Leiden. His research interests lie in the digitisation of textual transmission and reading, publishing studies, and scholarly communication. He is editor of a number of book series on these subjects, as well as of *Digital Humanities Quarterly*. His latest books are *Changing our Textual Minds: Towards a Digital Order of Knowledge* (2011), and *The Unbound Book* (2013), a collection of essays edited jointly with Joost Kircz. He is vice-chair of the COST Action 'E-READ', about the future of reading in the digital age, and is currently writing a book about reading.

Joris J. van Zundert is a senior researcher and developer in the field of digital and computational humanities. He works at the Huygens Institute for the History of The Netherlands, a research institute of The Netherlands Royal Academy of Arts and Sciences (KNAW). Joris J. van Zundert has headed and/or contributed to several key digital humanities projects at the Huygens Institute and the Royal Academy. He was chair for Interedition, a combined European USA network of digital humanities developers that fostered CLARIN in the Netherlands. As a researcher and developer his main interests lie with the possibilities of computational algorithms for the analysis of literary and historic texts, and the nature and properties of humanities information and data modelling. His current PhD research focuses on the interaction between computer science and humanities, and the tensions between hermeneutics and 'big data' approaches to interoperability and expertise exchange.

Claire Warwick is a professor in digital humanities at the Department of English Studies at the University of Durham. Her research is on the use of digital resources in the humanities and cultural heritage, on digital reading, and on the relationships between physical and digital information environments.

1. Introduction

Power, Practices, and the Gatekeepers of Humanistic Research in the Digital Age

Jennifer Edmond

This volume began, in many ways, with an image. A leaf floats downward from a tree and lands on the surface of a river below, from where it is carried away on eddies and ripples, to a new place far from its origin. There it may itself cause the formation of further rills and ripples in otherwise undisturbed water.

This image became a metaphor for how the emerging entanglement of technology and its imperatives with the practices and values of humanities research has become not only a point of intersection, but a roiling flow of both predictable and unexpected contingencies. The methodological moment in the digital humanities is well theorised and documented. At the level of the individual scholar, choosing to use, for example, a set of statistically determined topics generated by a software tool like Mallet, rather than a similar set extracted by the linear process of conventional reading, represents a difference in degree rather than kind. Like any methodological stake, the choice's implications for the knowledge it generates must be queried, understood, and accounted for in the scholarly claims that are based upon it.

But when the leaf becomes caught in the swell and passes downstream, the opening frame of its fall may be lost, and the leaf's path can be altered, just as many fallen leaves might later accrete to form barriers that may influence further flows. The sublimation of technology into the fabric of not just scholarly methods, but of the organisation of

© Jennifer Edmond, CC BY 4.0   https://doi.org/10.11647/OBP.0192.01

scholars and their work seems marked by a certain inevitability, not only because of the power of the methodological forces at play, but also because of the manner in which similar technologies are changing interaction and communication in the wider society.

The metaphor then became a meeting. In 2013, a very diverse set of stakeholders came together under the aegis of the ESF-funded Network for Digital Methods in the Arts and Humanities (NeDiMAH) in order to discuss how we, as policymakers, as librarians, as funding agencies, as humanists, and as computer scientists, might make sense of the changes technology was, and was not, bringing into our professional lives. The issues that this meeting raised also seemed diverse, even disconnected, at the time, and the measures that the stakeholders proposed to address them were equally multifarious and fragmented. After all, the tensions that technology introduces into research processes are more easily managed during certain phases than others: according to the principles of academic freedom, the scholar chooses her field of enquiry and can (within reason) define the appropriate methods for addressing her research questions. But the creation of knowledge is only part of the scholarly process; the sharing of that knowledge is an equally important part, a fact that can force a mismatch between the media and the message of scholarly communication.

The meeting focussed on these issues, and as a result produced two major outputs, the first of which was a useful taxonomy of objects that could be viewed as scholarship, including suggestions as to how they might be ideally disseminated and evaluated. This taxonomy divided the landscape of scholarly output that one might find in the digital humanities into six categories, only one of which has a clear precedent and place in the traditional flows of production, dissemination, and evaluation, namely print paradigm publications within closed formats (such as PDF documents). Alongside this we included electronic paradigm publications as a broad category that included everything from enhanced publications to blogs and Twitter corpora, to arguments presented in video and audio. Another paired set of entries in the list included single or collected/curated primary sources alongside datasets comprised of simpler objects, such as query results and intermediary processing files. Software was given a category of its own, as were methodological and teaching resources. We also considered patents/

licenses and ephemera (such as exhibitions and performances) as outputs, but concluded that the former category is more of a validation mechanism for other sorts of output, while the latter necessarily requires some documentation, making it ineligible to stand as a category in its own right.

The set of categories we devised made a powerful statement about the future of research, as only one of the six would be readily accepted in many evaluation contexts for many humanists — such as those applying for an academic position or building a case for promotion. For this reason, for each of the categories the group extended their work to include, firstly, the manner in which such work would be disseminated or communicated to an audience (a non-negotiable aspect for any research output, for research that is not ultimately shared with a community of practice cannot form the basis for further work); and, secondly, the basis upon which such work could build its credibility and be validated by the community. This latter category showed a large and interesting variability, encompassing well-embedded classifications such as peer review and citations, but, also, other forms of reuse, extensions in development, funding body review comments, uptake in training programmes, policy impact, community engagement, downloads, and even imitation.

In addition to the taxonomy, the 2013 group also developed a joint position paper aimed at finding common ground on the issues we observed from our various positions in the ecosystem. Neither of these outputs was ever formally released as an independent publication, though the work did instigate some discussion, especially among policymakers and funders struggling to define policies around the evaluation of digital scholarship. As with so many other discussions on the issue of where technology was taking the humanities, the work remained in the powerful, and yet denigrated realm of the informal, as many contributions to this volume will demonstrate. Indeed, this particular separation seems to be one of the primary axes of disruption within the humanities community wrought by the digital: precisely that the lines become blurred between informal communication and validated scholarship. The distinction between, for example, a position taken in personal correspondence and the line of argument in a scholarly monograph is easily distinguished; but the proliferation of forms — from the tweet to the blog post, to the listserv contribution, to the enhanced finding aid, as well as the public distribution and peer response inherent in many of these formats — has proven to be a test of our assumptions about where acts of scholarship lie, and what they consist of. As knowledge creation and consumption paradigms change, the authority that used to be reserved only for formal communications is shifting.

At the macro level, the growing acceptance of research approaches such as 'citizen science' has pointed toward this shift; but, even within circles of professional researchers, the nature and sources of scholarly authority have become contested. For example, a major issue identified at the 2013 NeDiMAH network meeting was the need to differentiate between two divergent processes: communication and publication. The difference between the two is defined here as: making your data and results public (communication); or, submitting them to peer review or other sort of verification by the scholarly community (publication), which may or may not include the publisher editing, enriching, and enhancing the work. In particular, this configuration (and the hierarchy it implies) is under pressures brought about by the emergence of 'impact' as a new value in scholarship, and the need to justify research spending in publicly-funded systems. The need for both of these processes is increasingly clear, but the relationship between them is increasingly muddy. The issues of evaluation and marketing are implicated here, as is the question of publication format and what to make available: form and content are both very much in play in the current environment, which creates particular challenges and opportunities.

This is not to say that all of the meta-discussion about how to assimilate the digital into the traditions of the humanistic epistemic culture has been informal, or even low profile. A 2011 special collection of articles about the evaluation of digital scholarship, edited by Susan Schreibman, Laura Mandell, and Stephen Olsen,[1] presented a very clear-eyed and practical roadmap for considering these issues. The LAIRAH survey[2] had already given us (as early as 2006) an empirical view of

---

[1] 'Evaluating Digital Scholarship', ed. by Susan Schreibman, Laura Mandell, and Stephen Olsen, *Profession* (2011), https://www.mlajournals.org/toc/prof/2011/1

[2] Claire Warwick et al., 'If You Build It Will They Come? The LAIRAH Study: Quantifying the Use of Online Resources in the Arts and Humanities through Statistical Analysis of User Log Data', *Literary and Linguist Computing*, 23.1 (2008),

the issues underlying how new types of scholarly object might be perceived and optimally matched to their users and environments. Far more recently, Smiljana Antonijević's large-scale ethnographic study of digital humanities (DH) and 'DHers' *Amongst Digital Humanists*[3] has done the same for the question of how skills and capacities are developed within emerging and traditional spaces and interactions of DH. And yet, despite all of the excellent work being done, it seems that the fundamental cultural change required to mainstream uniquely digital activities alongside digitised ones that are construed as surrogates for analogue processes, is still very much incomplete.

Since this discussion began, some things have changed, while others have not. The recognition that the digital is transforming research, including, and perhaps at this moment especially, in the humanities, is more widespread. That said, the critical traditions and strong commitment to qualitative approaches inherent in the humanities leave the digital humanities at risk of being caught between the poles of conservatism and technological disruption. This raises the question of whether the triangulation with digital methods changes the work of the humanities' disciplines only in degree, or indeed in kind. In spite of this, digital humanities is no longer merely a rare or niche approach that is fashionable yet suspect, but is rapidly becoming an embedded modality in the scholarly landscape. As a result, much of the growth in the impact of the digital on scholarship is now happening not so much 'at scale' in the large projects and research centres, but in the 'long tail', among researchers who might vehemently deny being in any way digital scholars, but whose work is still marked by the way in which technology transforms their interactions and interferes with the scholarly ecosystem. The manner in which the policy environment is embracing and encompassing the digital provides an assurance of this, albeit not necessarily a comforting one.

Take, for example, the development of the European Open Science Cloud (EOSC), an institution that, at the time of this volume's release, is still very much in flux, but also very much in the minds and mouths

---

85–102, https://doi.org/10.1093/llc/fqm045. The acronym LAIRAH stands for Log Analysis of Internet Resources in the Arts and Humanities.

3   Smiljana Antonijević, *Amongst Digital Humanists: An Ethnographic Study of Digital Knowledge Production* (London, New York: Palgrave Macmillan, 2015), https://doi.org/10.1057/9781137484185

of European research policy makers. The prospect of the requirement that all funded European researchers deposit their data in an open repository for reuse by others forces us to face a host of questions that would have otherwise lain unresolved. Who owns the source material that comprises the fundamental building blocks of research in disciplines like literature, history, music, or art history? And if the answer is that we, as researchers, do not own them, how are we to share them? What are the new data streams and sets that humanists create? Should paradata be more formally captured during the research process, and if so, how do we untangle it from the uniquely formed scholarly instrument of the individual humanist so as to make such data epistemically available to others? Indeed, what are humanities research data anyway: should this term be understood to encompass all inputs, outputs, and intermediary products related to our processes; or only those digital, quantifiable, relatively tidy streams and collections that are readily processed, federated, and aggregated? A recent Twitter thread initiated by Miriam Posner[4] illustrates yet again, that many humanists resist the term 'data' as a descriptor for their primary and secondary sources, or indeed for almost anything they produce in the course of their research. The fact that humanists already have a much richer and more nuanced vocabulary to describe these research objects is surely a part of the reason for this resistance, but the manner in which the term 'data' is deployed in disciplines that are primarily data-driven may also be a part of the hesitation concerning its adoption. In computer science, for example, this one word can be used to describe inputs, results, or intermediary research outputs; it can be relatively simple or highly complex; and it can be human-readable, or only intended for machines. The differentiation in provenance, value, importance, and authority of these different types of objects is one that humanists are highly sensitive to, making the adoption of the word 'data', with all of its slippery overdetermination, problematic indeed. It may also be that the lack of recognition of data as a humanistic object, outside of very narrow confines, has to do with the manner in which the practices of humanities research are differently institutionalised from other

---

4   Miriam Posner (miriamkp), 'Humanists out there, specifically non-digital humanists: If someone were to call the sources you use "data," what would your reaction be? If you don't consider your sources data, what makes them different?', 31 October 2018, 11:50 AM (tweet).

disciplines. For example, a lack of tools such as licenses and patents to protect intermediate or early stage findings drives a certain amount of caution in the readiness to release certain kinds of research output. If you cannot protect your knowledge capital at a granular level, then the potential to recognise these objects as elements in a category with a value and status of their own, diminishes.

The expectation implicit in the development of EOSC is, as one recent policy paper stated, that 'the researchers' job is based on data and on computational resources'.[5] However, very little humanities research is based on a single form or source of knowledge, with corroboration or triangulation between sources being more the norm. While big data research may be a rising paradigm across disciplines, humanists rarely value this form of foundation for research, seeing it as lacking a theoretical basis.[6] The digital does not just change the method, it changes the possibilities. The dark side of the digital humanities has always been its gravitational effect in pulling scholarship toward positivism, that is, towards the pursuit of research questions not because they provide insight into who we are as a species and where we have come from, but simply because the material to answer these questions is readily available and fits the tools and methods we have been able to borrow from elsewhere.

In this we return to the meta-level of organisation, the locus for sharing and valorising scholarship in the humanities in the digital age: what we are really speaking about here is power, and the shifting of power relations and conceptualisations of valid and invalid claims to epistemic agency. Péter Dávidházi, in the introduction to the volume he edited on changing scholarly publication practices, gives an interesting historical account of how this can operate.[7] Viewed on another temporal plane, I have written elsewhere about what I refer to

---

5 European Commission, *Prompting an EOSC in Practice: Final Report and Recommendations of the Commission 2nd High Level Expert Group on the European Open Science Cloud (EOSC)* (Luxembourg: Publications Office of the European Union, 2018), https://ec.europa.eu/info/sites/info/files/prompting_an_eosc_in_practice_0.pdf

6 Thomas Stodulka, Elisabeth Huber, and Jörg Lehmann, 'Report on Data, Knowledge Organisation and Epistemics', *KPLEX* (2018), https://kplexproject.files.wordpress.com/2018/06/k-plex_wp4_report-data-knowledge-organisation-epistemics.pdf

7 Péter Dávidházi, *New Publication Cultures in the Humanities: Exploring the Paradigm Shift* (Amsterdam: Amsterdam University Press, 2014), https://doi.org/10.26530/oapen_515678

as the 'generational fallacy',[8] an assumption according to which cultural change will inevitably result as new generations of scholars with a different relationship to technology enter the ranks of researchers. While it is true that those who are only starting their careers now may have a different level of comfort with technology and the kind of communication it fosters informing their assumptions and personal practices, what is obscured by underlying assumptions of this fallacy is the pervasive impact of the power relations and hierarchies, in particular in such a self-regulating system as a research community. Early career researchers may have excellent ideas for how to disrupt the system of scholarly communications, but if they also want to be successful in the fiercely competitive academic job market, they will have to make sure they do not try to push the paradigm too far or too fast, or they will risk alienating the most conservative reviewers of their work, be that on a key journal's editorial board, an appointment committee, or a promotion review board.

A discipline must maintain its ability to validate the work created within it, else it risks fracturing, and possibly even fissuring. This issue comes to a head in the evaluation process, but can also be seen, for example, in the discussion around whether humanists have data, or whether digital work lacks methodological rigour. Technology is not necessarily creating new points of pressure, but rather re-revealing old ones, such as disagreements about the differences between scholarship and service (a demarcation that impacts upon the credit given to coding humanists now, just as it has upon scholarly editors for decades, if not centuries), between important and average results, between quality measures and their proxies, and between the goals of our processes and the compromises we have negotiated to reach them in different times and under different conditions. Even the fact that the set of disciplines we so often refer to under the blanket term of 'the humanities' are themselves highly diverse — in terms of methodologies, in terms of expectations, and in terms of the availability and nature of sources — is implicated here. This is not so much a change in static, neutral processes (though it is that too) as a change in culture, in the values we promote,

---

8   Jennifer Edmond, 'OA, Career Progression and the Threat of the Generational Fallacy', *Open Insights Blog* (Open Library for the Humanities, 17 September 2018), https://www.openlibhums.org/news/304/

in the behaviours we tolerate, and in the language we refine to describe our experience as scholars. Not only do the current trends have deep roots, they also have a grounding in professional identities that are subconsciously cherished: facts that make them all the more difficult to resolve.

These tensions are not without their effect on the ecosystem as a whole. The system is riddled with markers of quality, prestige, and authority that are reliant on established proxies. Some of these, such as journal impact factors and citation analysis, not only create artificial demarcation of the places where quality scholarship can appear, but also, by their very nature, constrain the manner in which scholarship should be presented (which, for humanists, may in the first instance, mean books, chapters, and articles). Scholars must make a calculated decision when choosing to embark on a digital project. This decision not only concerns their research questions, their digital tools and methods, and how best to address and implement them; it also concerns their careers, their institutions, and their scholarly record. In spite of a general recognition of the value of digital scholarly outputs, many institutions and national systems still struggle to judge the merit of such outputs and to credit their creators accordingly. Interestingly, many of these trends, slow though they may be, point toward an even more fundamental disruption in scholarly communication, one that transcends the focus on output and products, to see scholarship instead as something living and evolving as processes. The idea that the work of scholarship needs to be 'fixed' before it can be evaluated is an essential aspect of our current system, one that is challenged by many aspects of the system we now see emerging: one of blogs as well as articles, open as well as double-blind reviews, and co-creation with citizens as well as unchallenged scholarly authority. But how can we trust what we cannot hold fast?

The manner in which trust is negotiated in the digital realm is an issue that reaches far beyond the question of how technology is changing the practices of humanities research. But it is most certainly another area where the foundation of our community consensus about the definition of scholarship, and how it acquires authority, is being tested. Print editions would have borne the authority of their authors as well as that of their publishers and editors who invested in them.

The digital edition may have no such proxy available, although many do prominently display the equivalent badges of the funder, project, or institution who sponsored them. Humanities research has largely been spared evaluation via blunt bibliometrics, as the data and instruments available are still ill-matched to the practices within these disciplines. But with the processual shift and the rise of alternative metrics, the question of how we can distinguish authoritative work and popular work adds new layers of complexity to these issues. Similarly, the use of sources of material that themselves may not have been validated in a formal or informal process, such as blogs or even Wikipedia, give rise to further concerns about what merits scholarly consideration and what does not.

The territory downstream from the digital humanities (and perhaps the territory of digital humanities as a whole) is, if nothing else, plagued by fragmentation: of institutions, of projects, of infrastructure models, and, indeed, even of the different understandings of what digital humanities and, more importantly, what digital scholarly communication flows in the humanities are or should be. Tensions in the wider research and publishing culture seem likely only to exacerbate this. For this reason, the authors of this volume believe that the work presented here is both timely and necessary, as both an attempt to create consensus across some of the existing boundaries and silos, but also to ring a warning bell for any of the systematic perversity we may be creating.

Scholarship does not arise in a vacuum, but rather within a complex ecosystem of ideas, people, structures, institutions, marks of esteem (like acceptance at a high-profile conference or invitation to sit on a board), and marks of negative judgement (like denial of promotion). In the current climate, many of the wider social drivers toward digital forms of communication and publication of and about scholarship come into direct conflict with the still dominant traditional modes of rewarding that scholarship. Many of our communal norms regarding quality are actually proxies that are dependent on the old model for their relevance, for example, journal impact factors only apply to journal articles, and publisher reputations only apply to books.

Within this system, institutions are beginning to recognise their own power to define new hallmarks for scholarly quality; but national and other pressures for comparability between institutions, and the continued persistence of the old heuristics within the community itself, do not necessarily support them in taking these courageous steps. Instead, we have seen the parallel development of a new rubric for evaluating knowledge creation, that is of impact, rather than excellence. However, many measures of impact, such as hit rates or media attention, are viewed as almost antithetical to the traditional norm of scholarly excellence of new knowledge being so rarefied as to be only accessible to other experts.

There is, of course, the question of how we counteract the insidious, transitional misgivings we still seem to have about digital sources not being 'real', and scholarship conducted in a virtual environment as somehow being less worthwhile because it is viewed as having been in some way 'easier' to create than via traditional modes of scholarship, which involve travel and discovery among dusty records without the assistance of Google Translate or our digital camera. Again, if we are to make progress in supporting the scholarship that is appropriate for our age and our disciplines, we will need to return to the primitives of knowledge creation and value those, rather than the romantic vision and symbolic authority of our accepted proxies.

Alongside the issue of how we understand scholarship in emergent formats is the concomitant issue of how we give credit for work done. The entrenched practice of quoting an original source, rather than the edition or digital facsimile you may actually have consulted, gives *short shrift* to both the hard work of scholarly editors, and to resources, particularly digital ones. With the current expansion of style guidelines to include citations for all sorts of works and formats, as well as tools like Zotero to make this process easier, there is no longer any reason for this complete misrepresentation of the point of access to research materials that we use. Both new and digital forms of scholarly output may need to include recommendations for users as to how the resource can be cited (be that in a monograph or within software code), but we also need authoritative confirmation of the importance of this practice. Should standards such as the MLA (Modern Language Association) style (or any other of the myriad options developed for specific disciplines) include a reference to a resource's site of access? How do we ensure we fully cite collaborative, non-traditional work? Do we need to reassess the demarcation between reference works (like bibliographies) and primary works? Primary and secondary works? How do we cite with the granularity of page numbers in a digital work (or does it matter anymore when we can simply issue the 'find' command)?

There is a lot of concern in the community about the reliability of digital scholarly outputs: after all, how do we evaluate, or indeed how do we even reference, what we cannot 'fix in place'? The guarantor of a book's durability is established in the institution of the library. The existence of multiple copies of a physical object (beginning from the point when the age of print was established) means, in the end, that these collections provide a relatively trustworthy, but perhaps less than systematic, guarantee that things held to be important in their own age will likely be available to future scholars somewhere when they need it. We have no such guarantees for the objects being created now, as neither libraries, universities, presses, research centres, or national agencies have a clear (funded) mandate to ensure these objects remain accessible in their current formats and in migrated formats into the future. This fear that resources could disappear, wholly or in part, diminishes the coinage of the digital output. Addressing this difficulty will be a part of the process of ensuring their equal status with traditional publications. Self-archiving is a good strategy for this in many cases, with copies maintained at institutional level, nationally, or by pan-European organisations, but this will have its limitations if there is a reliance on 'not for profits', lack of semantic encoding, or insufficient sophistication applied in archiving.

Of the many issues that intermingle and influence each other in this complex and fast-changing environment, three in particular — one 'upstream' and two definitively 'downstream' — merit a further detailed introduction. Each of these represents a paradigm in which identities, positionalities, and power hierarchies are either being exchanged or entrenched in the face of great change. These three factors are: the impact of collaboration as a mode of work in humanities scholarship, and the places of both publishers and evaluators as 'gatekeepers' for the acceptance of scholarship.

The Impact of Collaboration

We are not only moving toward a different paradigm of communication, but also toward different paradigms of knowledge creation, an additional shift that will have significant impact. Collaboration is a term that has come to mean many things in the current environment, from co-creation and co-authoring,[9] to the casual sharing of information and validation of others' results that has always occurred within scholarly communities. Knowledge sharing paradigms are perhaps still primarily imagined as unidirectional processes, flowing from expert to novice; but in reality, the complexity of the research questions being tackled today is such that knowledge is increasingly densely networked, partial, and reliant upon multiple intelligences in order to reach conclusions. This move toward greater integration between disciplines should not, indeed cannot, be forced (although it can, and needs to be, taught), but when it does occur it should be possible to validate and reward it. However, rewarding collaborative work is more than just an issue of deciding how much credit should go to how many people. Collaboration also brings a cross-fertilisation of methodologies, which is productive for enquiry, but creates tensions in a system where senior colleagues may be asked to evaluate the work of others whose epistemological frameworks have been defined according to a foreign idiom (critical theory, at least, was text — but software?). As such, the collaborations at the heart of the digital humanities tear at the fabric of the disciplines and many of the institutional structures that support and organise scholars and scholarship — hardly safe or solid ground. And the nature of these collaborations is not only interdisciplinary, but inter-sectoral. No one ever promoted an editor to full professor on the basis of their work on another author's book, and yet the importance of our collaborators across disciplines and sectors is growing so rapidly that the emergence of such a practice seems not just possible, but imminent. Nonetheless, there remains a deep discomfort in many places in the academy, even with co-authorship, in spite of its central role in supporting digital methodological approaches and their diverse outcomes. This stymies individual professional pathways, and also the development and

---

9   Joe Parent and Joe Uscinski, 'Of Coauthoring', *CRASSH* (19 June 2014), http://www.crassh.cam.ac.uk/blog/post/of-coauthoring

visibility of the digital humanities. A better understanding of what the various actors in the system, including potential industry and non-academic partners, 'want' and what they 'do', would go some distance to addressing these inconsistencies.

Until we can see coding as a generic capacity like reading or writing, the collaborative model of the digital humanities is likely to endure. But how does this become something that can transcend the power structures and the pitfalls between the disciplines? The ideal would be to see research questions and collaborations negotiated on the basis of reciprocity, that is, a relationship where each researcher brings their own questions to a given trajectory of research, and in which humanistic questions are pursued in concert with an advancing baseline of technological capacity. Until we are all fully 'multilingual' as pertains to technology, we will continue to need translators; but within a research context where the baseline assumptions and strengths of the convergent disciplines are so different, it does not make sense to view these individuals as lacking epistemic impact. Digital humanities work cannot be based upon the maxim of 'garbage in, gospel out'. Just as the precondition for the use of any text-based methodology would be that one read and understand the critical, theoretical, or methodological material being applied; the precondition for the application of digital tools must be that a scholar understands how they work and what they can be used for.

Evaluators as Gatekeepers

A further area of downstream concern for the digital humanities is that of how one evaluates the scholarly quality of these non-traditional publications, and traces their impact. Not everything produced by a scholar is a work of scholarship, and not everything produced within the digital humanities is of equal quality. Funding agencies and university departments alike are struggling to reimagine their evaluation processes, and are becoming less reliant on their own ability to see and judge the merit of their colleagues' work on a comparative basis with their own, and are instead investigating opportunities for accepting and evaluating the quality and impact of the work on its own terms. Even citation norms, which generally see researchers citing an 'original' print edition,[10] even if the work was based largely on digital surrogates, represent an ineffectual transfer of analogue habits to a digital context. While it may be seen by the individual scholar as irrelevant how exactly they reference their work, this ineffectual transfer may hide the potential contribution of the digital edition, and the possible impact of its construction and organisation on the trajectory of further investigations based upon them.

This crisis of conscience in scholarly evaluation hits the digital humanities particularly hard: the catch-22 of the new forms of scholarly output is that one wants to feel assured that one's work will be recognised, but that recognition is generally contingent on a certain familiarity and critical mass of accepted examples. Early adopters applying digital methods are at times 'punished' for making this choice by being required to write a traditional interpretive essay to accompany their digital work (with this essay being the only part of the output actively evaluated). A renewed requirement for deepening our understanding of what we expect from scholarship is created not just by new methods, but by the new objects produced by scholars in the digital age: books, journals, blogs, collaborative texts (wikis), databases, algorithms, software, coding, maps, images, 3D models and visualisations, videos, schemas, and documentation. The old proxies of press and journal reputations will not assist us in appreciating these highly influential new forms of scholarly communication: so, a part of the solution must lie in an enhanced need for explicit methodologies, which are documented and, therefore verifiable. All too often, technology, once applied to a problem, retreats into a 'black box' and fades from the discussion. This, however, undercuts the desire for rigorous, repeatable scholarship. The ideal scholarly output would allow others to manipulate the same data and to verify a colleague's results, or to produce new knowledge with the same data. This would be a realisation of the trend, discussed above, to reposition the end goal of scholarship from a fixed product to an evolving process, but the expectation that this could happen easily would be naive, as it is the nature of the humanities scholar to build his

---

10 Jonathan Blaney and Judith Siefring, 'A Culture of Non-Citation: Assessing the Digital Impact of British History Online and the Early English Books Online Text Creation Partnership', *Digital Humanities Quarterly*, 11.1 (2017), http://www.digitalhumanities.org/dhq/vol/11/1/000282/000282.html

or her personalised epistemic instrument on the basis of a long process of curating and assimilating resources and influences. This fact, which makes it difficult to step into the process of another scholar, or even to reuse of their data, is something we struggle to adapt to.

Publishers as Gatekeepers

The role of the scholarly publisher, traditionally our primary gatekeeper for the validation and production of scholarly resources, is splintering. The physical production of tangible book objects is only a small part of the process, so the reduction in importance of this stage in the process alone does not in any way mean that all points in the chain from author to market are being adequately covered by the new landscape. The acceptance process was, and still is, a powerful marker of perceived quality, a proxy upon which we seem reliant, in spite of our slightly bad consciences about it. The editing function and rights clearance must also still exist. The creation of a durable object is easy with a book, and much harder with a web publication, a tool, or piece of software. The marketing and selling functions also should not be underestimated as being part of scholarly dissemination, in particular as audiences are becoming multiple and varied: from the small community of specialists, to works of vast, popular, as well as scientific, interest. Finally, with the democratisation of publishing itself, came also a raft of difficulties with understanding who was reading what and why. Usage metrics are complex and often flawed, in part because what we know (and what we need to know) about reading books is not comparable to what we know (and need to know) about reading online. In an ecosystem where traditional publishers (with and without their own online presence) and new open access (OA) publishers coexist with independent peer reviewers, self-publishers (from individuals to universities), and everything in between; a new understanding of the scholarly communication's 'value chain' and the best practice for forging all of its links is a fundamental requirement. This new understanding should be able to encompass all forms of publishing, from the traditional to the avant-garde, utilising the strengths and mitigating the weaknesses of each.

All too often, the discussion about the emerging role and responsibilities of these particular gatekeepers becomes overdetermined by concerns of the cost of providing access to scholarly materials. More and more scholarly materials are now available online (whether created as a digital native object or not), and some research methods (such as those based on data-mining techniques) and collaborative relationships are contingent and reliant upon this availability. Furthermore, even within a largely digital ecosystem, less established researchers, or researchers from less affluent countries or institutions, may have substantially less access to material. It is therefore of the greatest benefit, from a researcher's perspective, to have them as widely accessible as possible. Open access does not mean free, only free at point of access, and key elements of this development would be to create business models for this mode of publishing that fit the humanities' publishing practices (such as print on demand for monographs, for example). We need, as well, to understand when openness is inappropriate, for example, in cases where copyright or confidentiality may prevent any publication if open access is the only option. There are both ethical and economic arguments for the provision of greater access to scholarship, but we also need to be wary of the turning of the current discussion to article processing charges (APCs) as a solution to the imperatives to provide wider access to scholarship: while this might ease the situation on the user's side, we could easily create a different risk, that is, that publication in the best journals will become tied to the author's ability to pay, rather than to the quality of the scholarship only. The 'green/gold' debate around open access to research outputs has focussed a lot of attention on this part of the pipeline, but it is important to be aware of the potentially perverse incentives this focus might bring. Underlying it are, for example, assumptions around access to funding and/or that the best research takes place in the context of an externally-funded project. While the humanities will be required to respond to the wider trends in research policy, it is important to make sure that the core values of the research, along with the value of the research itself, is protected, even as the social contract with its gatekeepers is being actively revised.

However, access is an issue that goes beyond the parameters of the debate around the deposit of scholarly research with trusted public or institutional repositories. Access to materials also encompasses issues of conservation (for it is to the analogue originals that many people want access, with the digital surrogate being just that, a surrogate), and linguistic availability to scholars who may not have mastery over the language of a particular discourse. While these issues may be beyond the reach of a project with its basis in digital methods, their impact must be recognised and incorporated (if only at a background level) into any discussion of humanistic scholarly communications in the digital, or any age.

Finally, there are macro-level issues surrounding the technical and legal frameworks for sharing the output of digital humanities projects. How can we be sure that individual works of digital scholarship will be available in the long term? How can we reimagine issues of copyright and 'fair use' so as to enable the kind of deep citation and linking these projects might utilise? While these debates extend in their scope from the divergent copyright laws found in individual nations, all the way down to the preservation mandates of universities, they still must be recognised as significant, potential barriers to the widespread uptake and mainstreaming of digital humanities' methods. As the role of the publisher changes, our traditional partnership in the negotiation of these issues may deteriorate.

This Volume's Contribution

The chapters in this volume are perhaps not so much about scholarship as they are about the scholars who create them and the manner in which they negotiate the relationships and flows of knowledge that pass between them. It is, after all, people and the systems around them that decide what is and is not a meaningful contribution to knowledge. Some of these contributions date back to the time of the NeDiMAH network meeting, and, though they have been updated, the issues they raise still seem astonishingly fresh. Other contributions respond to some of the latest trends in the research environment and how the issues expressed in this introduction are being stymied or promoted by wider trends in research policy and scholarly communications.

In general, this volume can be seen as consisting of discursive pairs of contributions (although the authors of the individual chapters are not necessarily responding directly to each other's work). The Chapters

1 and 2 look at traditional publishing models, the functions they serve, and the changes occurring in how they act as gatekeepers for scholarship. Focus then moves in Chapters 3 and 4 to the question of the validation of scholarship as seen through the lenses of both impact and scholarship as a market. The Chapter 5 looks at disruptions and continuities in specific forms of research practice, exploring in particular the narrative argument in codework. The next pairing, Chapters 6 and 7, delves into the history of our discussion of these changes, exploring early evidence for how we might evaluate digital scholarship in the humanities, and how emerging venues for scholarly communication come to be associated with certain kinds of validation and certain points on the continuum between formal and informal communications. Finally, Chapters 9 and 10 take a macro-level perspective and look at changing practices through the lenses of two emerging trends driven by European research policy: first, the development of bespoke research infrastructures for the arts and humanities, and second, the acceptance of the paradigm of FAIR (or 'findable, accessible, interoperable, and reusable') data, and its applicability to the humanities.

Through these various explorations, this volume sheds significant new light on the shifting practices in humanities research, which have been facilitated by technology but driven by a far wider range of impulses from scholars and scholarship. From product to process, from formal to informal, from published to communicated, these pieces delve into the shifts that many of us take for granted, exploring the impact they are developing on our work and identities as scholars. They prove that humanists not only welcome technology, but take ownership of it in unexpected ways. As such, it contributes not only to our meta-understanding of our work and world, but also empowers us to make a case for what form our scholarship takes, whatever it may be.

Bibliography

Antonijević, Smiljana, *Amongst Digital Humanists: An Ethnographic Study of Digital Knowledge Production* (London, New York: Palgrave Macmillan, 2015), https://doi.org/10.1057/9781137484185

Blaney, Jonathan, and Judith Siefring, 'A Culture of Non-Citation: Assessing the Digital Impact of British History Online and the Early English Books Online

Text Creation Partnership', *Digital Humanities Quarterly*, 11.1 (2017), http://www.digitalhumanities.org/dhq/vol/11/1/000282/000282.html

Dávidházi, Péter, *New Publication Cultures in the Humanities: Exploring the Paradigm Shift* (Amsterdam: Amsterdam University Press, 2014), https://doi.org/10.26530/oapen_515678

Edmond, Jennifer, 'OA, Career Progression and the Threat of the Generational Fallacy', *Open Insights Blog* (Open Library for the Humanities, 17 September 2018), https://www.openlibhums.org/news/304/

European Commission, *Prompting an EOSC in Practice: Final Report and Recommendations of the Commission 2nd High Level Expert Group on the European Open Science Cloud (EOSC)* (Luxembourg: Publications Office of the European Union, 2018), https://ec.europa.eu/info/sites/info/files/prompting_an_eosc_in_practice_0.pdf

Parent, Joe and Joe Uscinski, 'Of Coauthoring', *CRASSH* (19 June 2014), http://www.crassh.cam.ac.uk/blog/post/of-coauthoring

Posner, Miriam (miriamkp), 'Humanists out there, specifically non-digital humanists: If someone were to call the sources you use "data," what would your reaction be? If you don't consider your sources data, what makes them different?', 31 October 2018, 11:50 AM (tweet).

Schreibman, Susan, Laura Mandell, and Stephen Olsen, eds., 'Evaluating Digital Scholarship', *Profession* (2011), https://www.mlajournals.org/toc/prof/2011/1

Stodulka, Thomas, Elisabeth Huber, and Jörg Lehmann, 'Report on Data, Knowledge Organisation and Epistemics', *KPLEX* (2018), https://kplexproject.files.wordpress.com/2018/06/k-plex_wp4_report-data-knowledge-organisation-epistemics.pdf

Warwick, Claire, et al., 'If You Build It Will They Come? The LAIRAH Study: Quantifying the Use of Online Resources in the Arts and Humanities through Statistical Analysis of User Log Data', *Literary and Linguistic Computing*, 23.1 (2008), 85–102, https://doi.org/10.1093/llc/fqm045

2. Publishing in the Digital Humanities

The Treacle of the Academic Tradition

Adriaan van der Weel and Fleur Praal

Digital humanities (DH) scholars use novel digital tools and methods to help answer research questions that are difficult to handle without the aid of a computer. Sometimes, too, these new methods and digital tools profoundly reshape the very nature of the questions themselves. Moreover, the need for the continuing development of state-of-the-art technology adds a problem-solving dimension to the research.[1] Taken together, these characteristics justify the sense that DH is not just a divergent scholarly field, but even a disruptive one.[2] Given DH's

---

1   The sociology of science aims to explain research and communication practices in particular academic fields by modelling their research objects, methods, and approaches (epistemology) in a multidimensional classification. DH can be said to diverge from more traditional humanist disciplines by accommodating greater external influences in research application, and by constituting a technology-driven research front. For a full exposition of such theories and classification models, see: Richard Whitley, *The Intellectual and Social Organization of the Sciences* (Oxford: Oxford University Press, 1989); Tony Becher and Paul Trowler, *Academic Tribes and Territories: Intellectual Enquiry and the Culture of Disciplines* (Buckingham: Society for Research into Higher Education & Open University Press, 2001); Wolfgang Kaltenbrunner, 'Reflexive Inertia: Reinventing Scholarship through Digital Practices' (unpublished doctoral dissertation, Leiden University, Leiden, 2015), http://hdl.handle.net/1887/33061

2   We will refer to the various geographically distributed communities and methodological specialisms in DH as belonging to one disruptive movement, in comparison to the traditional research fields in the humanities. This by no means serves to argue that DH would be a homogeneous field: we are fully aware that beliefs and practices vary across language communities, subject domains, and

wide and eloquent conceptual support for the use of novel tools and approaches to humanist knowledge construction for all purposes, one would expect such a field to employ innovative communication practices as well. Indeed DH projects, probably to a greater extent than is the case in the more traditional humanities fields, are often communicated through databases, websites, datasets, software tools, online collections, and other informal means of making results public.[3] However, while DH is clearly taking on a pioneering role in experimenting with such new communication forms, there is a problem when it comes to their recognition as formal publications. Even where these new digital-born forms of research output may communicate knowledge that is just as valuable as that found in traditional print-based publications, they still do not achieve similar authority. They are not generally regarded by tenure committees and funding bodies as the equivalents of formal scholarly articles and books, and scholars do not rely on them as heavily or as frequently as on formal publications, or at least do not acknowledge it as confidently. In consequence, when all is told, DH *publication* practices — as distinct from communication practices at large — diverge less from mainstream practices than expected.

If the impact of experimentation in DH on publication habits remains limited, what are the factors that inhibit the field's disruptive potential? In this chapter, we want to explore the discrepancy between the novel communication opportunities offered by new types of scholarly output, and the strong adherence to traditional, formal publication habits that persist even in an innovative community of practice such as DH. We start by arguing that books and articles occupy their particular position because of four functions of formal publishing that are

---

disciplines. Nevertheless, the observation that DH groups share more ideologies and communication routines with each other than with the traditional humanities fields legitimises our comprehensive description of them as an inclusive community of practices, as does the fact that a diverse, international and interdisciplinary population of scholars identifies themselves as belonging to the DH community. See also: Matthew G. Kirschenbaum, 'What is Digital Humanities and What's It Doing in English Departments?', in *Debates in the Digital Humanities*, ed. by Matthew K. Gold (Minneapolis: University of Minnesota Press, 2012), pp. 3–11, https://doi.org/10.5749/minnesota/9780816677948.003.0001; Anabel Quan-Haase, Kim Martin, and Lori McCay-Peet, 'Networks of Digital Humanities Scholars: The Informational and Social Uses and Gratifications of Twitter', *Big Data & Society* (2015), 1–12 (pp. 1–2), https://doi.org/10.1177/2053951715589417

3   See the comprehensive overview at https://eadh.org/projects

the — print-based — embodiment of fundamental academic values. DH cannot behave as if it were an island governed by its own laws. This explains why the acceptance of novel digital communication forms as authoritative scholarly output is much slower than technological innovation would justify, in academia in general, but even in a progressive and pioneering field as DH. Second, we will use this framework of the functions of publishing to analyse how the inherent properties of the new digital medium are beginning to challenge and destabilise paper-based conventions.

Is the adherence to convention in the DH community really as strong as we have suggested? In the following pages we will maintain the distinction we began to make at the outset between scholarly *communication* (the superordinate term, which includes all forms of communication and making public, both informally and through established publishers' channels), and the much smaller subclass of formal academic *publication*.[4] To begin with the former, we have already observed that the DH field is experimenting with a wide variety of means to disseminate research outcomes. However, even the communication habits of DH scholars are, perhaps, not as revolutionary as is sometimes claimed. Although it has, for instance, often been remarked that DH communities use Twitter intensively[5] — such observations have even been made by journalists attending DH conferences[6] — the scant analysis available has demonstrated that DH-Twitterers use the platform for discipline-relevant, research-related messages proportionally less than users from other fields.[7]

---

4   Fleur Praal and Adriaan van der Weel, 'Taming the Digital Wilds: How to Find Authority in a Digital Publication Paradigm', *TXT*, 4 (2016), 97–102 (pp. 97–98), https://openaccess.leidenuniv.nl/bitstream/handle/1887/42724/PraalvdWeel.pdf
5   Martin Grandjean, 'A Social Network Analysis of Twitter: Mapping the Digital Humanities Community', *Cogent Arts & Humanities*, 3.1 (2016), 1171458, https://doi.org/10.1080/23311983.2016.1171458
6   Kirschenbaum, 'What is Digital Humanities', 7–8. Kirschenbaum here puts observations by *The Chronicle of Higher Education* and *Inside Higher Ed* in context.
7   Kim Holmberg and Mike Thelwall, 'Disciplinary Differences in Twitter Scholarly Communication', *Scientometrics*, 101.2 (2014), 1027–42, https://doi.org/10.1007/s11192-014-1229-3. Holmberg and Thelwall identify a large group of DH-Twitterers who send more messages than the comparable user bases from other academic disciplines — but fewer than average of those messages indicate a clear link with scholarly activity. To our knowledge, there is no comparable research of a more recent date; Grandjean does not analyse tweet content, but focuses on the connections between users instead ('Social Network Analysis').

Especially where formal academic publication is concerned, DH practices turn out to be quite conventional. For example, it may well be true that DH engage in more intensive collaboration than the traditional humanities at large. However, this concerns, in particular, the pre-publication phases of research. Research projects often *require* collaboration, for example, because external technical expertise may need to be brought in, or because the creation of sufficiently large data sets requires the input of more than one person. However, when it comes to *publication*, explorative studies do not demonstrate a significantly increased occurrence of co-authored papers, and no increase in the average number of authors collaborating on book chapters.[8] In the meantime, the number of publications that attempt to define, explicate, and seek support for new research communication practices for DH is so large that it constitutes a veritable genre in its own right. Indeed the genre has often been cited as evidence of the reflexive tendency of the field.[9] Some argue that what makes the field of DH revolutionary in nature is its grounding in 'online values' that are fundamentally different to the norms of print.[10] Ironically, though, most of the publications in

---

[8] An analysis of Flemish humanities publications does not yield conclusive evidence; nor does a more recent analysis of two DH journals. Truyken L. B. Ossenblok, Frederik Verleysen, and Tim C. E. Engels, 'Co-authorship of Journal Articles and Book Chapters in the Social Sciences and Humanities (2000–2010)', *Journal of the American Society for Information Science and Technology*, 65.5 (2014), 882–97, https://doi.org/10.1002/asi.23015, http://onlinelibrary.wiley.com/doi/10.1002/asi.23015/abstract; Julianne Nyhan and Oliver Duke Williams, 'Joint and Multi-Authored Publication Patterns in the Digital Humanities', *Literary and Linguistic Computing*, 29.3 (2014), 387–99, https://doi.org/10.1093/llc/fqu018. We know of no other comparable research of a more recent date.

[9] Kirschenbaum lists many of the formative texts; from him, we have also borrowed the notion of classifying this ongoing discourse as a genre. Kirschenbaum, 'What is Digital Humanities', 3.

[10] Lorna M. Hughes, Panos Constantopoulos and Costis Dallas, 'Digital Methods in the Humanities: Understanding and Describing their Use across the Disciplines', in *A New Companion to Digital Humanities*, ed. by Susan Schreibman, Ray Siemens, and John Unsworth (London: Wiley Blackwell, 2015), pp. 150–70, https://doi.org/10.1002/9781118680605.ch11. Attributing great idealism and revolutionary fervour to the field is perhaps tempting, but it might be more constructive to regard the abundance of reflection as typical of any emerging discipline. These texts are the record of a community's attempts to modify the existing conventions of research and research communication. DH scholars' uptake of new communication technologies perhaps challenges the monopoly of print, but this challenge is not exclusive to the DH field. Furthermore, the challenge does not by itself revolutionise communication habits, it merely reinforces the need for adjustments.

this genre appear in conventional academic publications: articles or book chapters.[11]

DH — rightly so — continues to subscribe to the argument that new communication types should be acknowledged as valuable contributions to the scholarly endeavour.[12] Why, then, is a DH revolution in publication practices not happening? Why do the publication habits of such a youthful and unruly field still remain firmly grounded in the print-based paradigm? This paradox warrants a dispassionate appraisal of the communication and publication issues that confront DH. To explain why formal publication is especially slow to change, despite ongoing shifts in scholarly communication in general, we first examine the framework of established functions of academic publishing, and then contrast this framework with the inherent properties of the novel digital communication and publication technologies. In doing this, we will adopt the perspective of the scholarly author as a primary stakeholder actively steering through the myriad of available options.

The Functions of Scholarly Publishing in the Print Paradigm

In varying proportions, and depending on the discipline, monographs and articles in edited volumes and journals have come to constitute the narrow range of widely accepted formal academic publications. These are the designated text types of formal communication between peers in

---

11 James P. Purdy and Joyce R. Walker, 'Valuing Digital Scholarship: Exploring the Changing Realities of Intellectual Work', *Profession* (2010), 177–95, https://doi.org/10.1632/prof.2010.2010.1.177; Lisa Spiro, '"This is Why We Fight": Defining the Values of the Digital Humanities', in *Debates in the Digital Humanities*, ed. by Matthew K. Gold (Minneapolis: University of Minnesota Press, 2012), pp. 16–36, https://doi.org/10.5749/minnesota/9780816677948.003.0003

12 Bethany Nowviskie, 'Where Credit is Due: Preconditions for the Evaluation of Collaborative Digital Scholarship', *Profession* (2011), 169–81, https://doi.org/10.1632/prof.2011.2011.1.169; Jennifer Edmond, 'Collaboration and Infrastructure', in *A New Companion to Digital Humanities*, ed. by Susan Schreibman, Ray Siemens, and John Unsworth (London: Wiley Blackwell, 2015), pp. 54–66, https://doi.org/10.1002/9781118680605.ch4; Smiljana Antonijević and Ellysa Stern Cahoy, 'Researcher as Bricoleur: Contextualizing Humanists' Digital Workflows', *Digital Humanities Quarterly*, 12.3 (2018), http://www.digitalhumanities.org/dhq/vol/12/3/000399/000399.html

science and scholarship. A rich variety of other forms of communication exist in which academia has always created connections, discussed research findings, and generated new ideas — but they have been consistently branded as informal exchanges. As a result of the symbiotic development of print culture and the systemic values of scholarship over the course of four centuries, books and articles have been established as the gold standard of formal academic publication. Although these values are rarely made explicit, there is broad consensus that formal contributions to knowledge should be original; they should be made available for the academic community independent of authors' social standing; they should not serve any interest other than the furthering of knowledge; and they should be able to withstand systematic scrutiny.[13] Academics who uphold these norms can be esteemed for making valuable contributions to knowledge. Implicitly or explicitly, authors will seek to adhere to those values each time they communicate research results publicly. These values are enshrined in the four commonly identified functions of academic publishing: registration, certification, dissemination, and archiving.[14]

Dissemination is perhaps the most obvious goal, defined as the transfer of knowledge to others by 'making it public'. This does not happen indiscriminately; there is a strategic component to it. Both scholars and publishers strive to distribute texts among their optimal audience. Authors strategically select a venue for publication that ensures the widest possible distribution among the — often very small — group of experts they wish to target. Publishers filter the texts submitted to them on topicality and currency, and to suit the interest of a relevant and identifiable market to which they have — or seek to gain — access.

---

13   These are the values of Communalism, Universality, Disinterestedness, Originality, and Scepticism (CUDOS), first codified by Robert Merton and developed by John Ziman. Robert K. Merton, *The Sociology of Science: Theoretical and Empirical Investigations*, ed. by Norman W. Storer (Chicago: Chicago University Press, 1973); John Ziman, *Real Science: What It Is and What It Means* (Cambridge: Cambridge University Press, 2000).

14   H. E. Roosendaal and P. A. Th. M. Geurts, 'Forces and Functions in Scientific Communication: An Analysis of their Interplay', unpublished conference paper at *Cooperative Research Information Systems in Physics*, Oldenburg, Germany, 31 August–4 September 1997, www.physik.uni-oldenburg.de/conferences/crisp97/roosendaal.html; David C. Prosser, 'Researchers and Scholarly Communications: An Evolving Interdependency', in *The Future of Scholarly Communication*, ed. by Deborah Shorley and Michael Jubb (London: Facet, 2013), pp. 39–49, https://doi.org/10.29085/9781856049610.005

Second, publishing serves the function of *registration*: through publication, an author is acknowledged as the original discoverer, explicator, or analyst of the research object, and, in the humanist disciplines especially, also as the creator of the scholarly argument that describes the findings (i.e. the text that constitutes the publication itself). Published texts thus form the records of research, and demonstrate their originality as knowledge contributions. Engrained notions about authorship and the attendant esteem of 'being published', within and outside academia, stem from this registration function.

The esteem of authorship is also intimately connected with the function of *certification*. This is the legitimisation and crediting of the authors' claims through organised scrutiny during the process of publication. Editors and publishers filter submitted texts based on quality, topicality, and currency; the selected texts then go through a vetting process (and, often, subsequent rounds of revision) before they are published. This review mechanism is crucial to the way formal communication proceeds along the chain of stakeholders. Readers are aware that review happens, and select their reading based on assumptions about quality control; authors are aware that readers value scrutinised texts and, therefore, aim to publish in channels known for their rigour; and publishers depend on authors' and readers' awareness, to maintain their role as independent agents establishing credibility for scholarly communication.

Archiving, lastly, is the preservation of research within dependable systems to ensure that future generations of scholars will be able to build on existing knowledge. Libraries, with their book repositories and journal collections, grew to become publishing's chief archiving infrastructure. That their search and discovery systems are finely tuned to publication metadata forms an additional incentive for authors to publish a text formally, instead of only circulating it informally.

The system of scholarly publishing has come to rely squarely on the combination of these four different functions. Nevertheless, the different stakeholders in scholarly communication have diverging interests in the balance between those functions in every communicative instance. For example, 'a document that allows for a means of conferring reputation on a researcher may not be the same as a document that transmits the maximum amount of information'.[15]

---

15   Prosser, 'Researchers and Scholarly Communications', pp. 43–44.

Furthermore, we argue that even scholarly authors themselves, our primary stakeholders in this analysis, do not form a stable and homogeneous group. They demonstrate dynamic and contrasting mixes of priorities in their communication practices. They are aware — if perhaps only intuitively — of the functions of publishing. In general, this can be explained by the fact that all authors also act as scholarly readers, and therefore switch between these two roles and prioritisations.[16] Moreover, DH scholars are particularly prone to reflecting on their own practices as a direct extrapolation of their research topic, and they may be expected to provide more explicit reasoning for their choices.

Less visibly, this set of historically grown functions of publishing, in turn, largely depends on salient properties of the print medium. These properties constitute the technological and cultural frameworks in which academic publishing developed and that have come, over time, to be observed as a matter of course in the process of formal publication. They include, for instance, the assumption of the finality and fixity of the printed text, and its inherent duo-modality of text and images, but also 'the restriction to a predominantly textual format, only supplemented by the occasional use of graphs and charts or still images; the use of a rhetorically formal — even formulaic — and discipline-specific register; and adherence to a formalised and strictly methodical referencing practice'.[17] Therefore, academia — perhaps unintentionally — relies on the formal functions of scholarly publishing for inferring the value of a text. These formal functions in turn depend on largely implicit assumptions about the connection between the scholarly importance of a text and the properties of print.

---

16 For further observations on the varying — and even opposing — interests of the scholar-as-author and the scholar-as-reader, see the themed issue of *Against the Grain* on the future of the scholarly monograph, and in particular Adriaan van der Weel and Colleen Campbell, 'Perspectives on the Future of the Monograph', *Against the Grain*, 28.3 (June 2016), 1, 10, http://www.against-the-grain.com/wp-content/uploads/2016/07/ATG_v28-3.pdf

17 Praal and Van der Weel, 'Taming the Digital Wilds', p. 98.

Transferring the Functions of Publishing to the Digital Medium

Compared with the established printed forms of publication (chiefly books and articles), the digital medium affords new, and in some cases, very different possibilities. These can be explained by a number of properties inherent in digital technology that together can be said to characterise the medium. Just as the printed book was fundamentally characterised by materiality and fixity, digital technology in its online form can be said to be characterised by immateriality and fluidity, two-way linking, machine-readability (i.e. searchability), and multimodality. These salient properties and their affordances have major repercussions also for academic publishing.[18] Some of the changes it has brought to textual dissemination can be easily observed; for instance, online creation, lossless copying, and digital dissemination of content have allowed a decrease in production and distribution costs, while increasing the speed of these processes. The architectural flatness of the Internet gave rise to Web 2.0 networks characterised by a new interactivity in which, moreover, all data types converge. However, the wider but not necessarily intentional implications of the digital medium's salient features manifest themselves fully only gradually in the social reception of the technology. In the case of scholarly communication and publication practices, the rise of open access — which is predicated on the salient feature of lossless copying at virtually zero incremental cost — is a current example. Just as occurred in the case of print, technological invention is thus followed by a much slower sociocultural process of discovery in which the new medium's properties begin to influence actual communication practices.

As new tools and methods are developed in an increasingly quick succession of innovations, the digital medium's properties continue to affect research practice. Similarly, the evolving affordances of the online environment shape scholarly authors' expectations about communication. In this process of discovery, authors conceptualise

---

18 For a more detailed discussion of the role of inherent salient properties of textual media, see Adriaan van der Weel, *Changing Our Textual Minds: Towards a Digital Order of Knowledge* (Manchester: Manchester University Press, 2011), especially Chapter 3, 'The Order of the Book', and Chapter 5, 'Salient Features of Digital Textuality'.

the audience's response to their messages; and, in their turn, readers' expectations are influenced by prior experience in similar communicative situations. The sociotechnical adoption of any technological innovation is thus a complex system in which recurrent feedback loops drive change. The adoption of the online medium for scholarly communication leads to very gradual, iterative shifts in the norms and values of academia. Since authors are likely to desire faster change than readers, they are also likely to experience greater frustration with this slowness.[19]

This acculturation process has only begun recently, and normative change cannot yet be clearly discerned. Rather, the possibilities of online communication are initially embraced by authors in order to adhere, as much as possible, to the established functions of publishing — even if they will increasingly point to imperfections inherent in the print paradigm. Authors who are keenly aware of the online affordances and are willing to experiment with digital communication, such as is typical in DH, may be considered a progressive influence, potentially accelerating the processes of change. The research evaluation systems that science policy relies on, such as the British Research Excellence Framework (REF), or the Dutch Standard Evaluation Protocol (SEP), and their equivalents across the globe, on the other hand, *inherently* reflect existing practices and therefore reinforce established norms, and can thus be seen as conservative forces in the system. They make scholars conservatively opt for communication through acknowledged formal text types. However, as a result of the myopia with which these systems still connect books and articles with academic prestige and reward, they may also indirectly render academia more aware of the undesirable aspects of the dominance of formal publications in research, fuelling ongoing debates and experimental excursions.

The previous paragraphs have sketched the changing landscape of scholarly communication and publishing in broad strokes. In the following sections, we will engage in a structured exploration of current scholarly communication practices, situated within the established framework of the functions of publication as described above. Examples of emergent digital practices, as observed in the digital humanities or other directly relevant disciplines, point to conceptually shifting undercurrents in the value system of academia: today's online

---

19   See note 16 above.

experiments may come to be considered as the good scholarship practices of the future.

Dissemination

With its near unlimited storage capacity, lossless copying, and low-cost options for file transfers, the digital medium has come to affect, directly and very visibly, the dissemination processes of formal publications, even if it has not fundamentally altered the traditional content types. In their current born-digital format — usually PDF, which mimics the lay-out of print — articles and books are indeed less costly to produce, and certainly much easier to copy and distribute widely. However, such formal publication formats, while being born digital, truly remain products of the print paradigm. Undergoing the exact same publication process as their print equivalents have long done, they continue to exhibit all four of the functions of publishing. For scholarly authors — our chief focus — the only change in the process is that the paper end-product might now be accompanied (or replaced) by a digital equivalent. Formal publications 'gone digital', therefore, are no more than a digital surrogate. They do not present an alternative to the traditional functions of publishing themselves, even though dissemination has become near-paperless.

To find evidence of real innovation caused by the shift in dissemination affordances, we should look beyond the immediate technological effects for signs of social change, which, as we have argued above, take longer to make their appearance. Although the formal content types of print culture still remain the standard for authors,[20] the traditional tools that facilitate dissemination — such as

---

20 Recent research suggests that humanities authors increasingly create non-traditional research output, such as websites and blogs (over sixty-five percent of authors create these), and datasets, visualisations, and digital collections (around thirty percent): Katrina Fenlon et al., 'Humanities Scholars and Library-Based Digital Publishing: New Forms of Publications, New Audiences, New Publishing Roles', *Journal of Scholarly Publishing*, 50.3 (2019), 159–82 (pp. 165–66), https://doi.org/10.3138/jsp.50.3.01. The same survey indicates that humanities scholars still refrain from creating and citing online communication forms, because they feel that print is valued higher by peers and evaluation bodies, and because print-based publications ensure a more stable and durable record (Fenlon et al., 'Humanities Scholars', 161–62). Other research confirms that significantly fewer authors are

library catalogues and publishers' content marketing through well-known channels — increasingly get bypassed in favour of alternative online technologies. About thirty-five percent of humanities scholars report favouring Google Scholar as their starting point for literature research. This is a larger proportion than those who initially turn to national or international catalogues, and discipline-specific publishers' platforms such as JSTOR.[21] Besides formal publications, Google Scholar features reports, self-published texts, and citations in its search results, whereas publishers' platforms can only retrieve indexed formal publications. Although this is presumably not the initial reason why scholars have shifted to generic search engines, the fact that informal content types get exposed next to formal publications might help a gradual acceptance that they represent a certain value.

Besides generic search engines, scholarly communication networks are rising as popular instruments for content dissemination. The overwhelming majority of researchers maintain profiles on ResearchGate, Mendeley, or, preferred more widely in the arts and humanities, Academia.edu and their non-commercial counterpart, Humanities Commons (HCommons), using the platforms to disseminate their own works and access those of others.[22] These new technologies are no longer in the metadata-based, hierarchical content-ordering mould of the traditional dissemination services; rather, they successfully use the

---

inclined to recognise any other forms of communication as equal to traditional publications; blogs and contributions to online conversations, especially, are seen as less important than publications (by eighty-five percent of survey respondents). However, about half of the respondents value created software equally as highly as traditional publications'; this should be 'However, about half of the respondents value created software equally high [or: 'as highly'] as traditional publications: Christine Wolff, Alisa B. Rod, and Roger C. Schonfeld, *UK Survey of Academics 2015*, Ithaka S+R | Jisc | RLUK ([n.p.], 2016), esp. p. 44, fig. 24, https://doi.org/10.18665/sr.282736

21 Wolff, Rod, and Schonfeld, *UK Survey*, pp. 10–15. This report does not investigate the rationale for such behaviour; however, users' preference for generic keyword searches and a dislike of advanced search options may be cues: Max Kemman, Martijn Kleppe, and Stef Scagliola, 'Just Google It: Digital Research Practices of Humanities Scholars', in *Proceedings of the Digital Humanities Congress 2012*, ed. by Clare Mills, Michael Pidd, and Esther Ward (Sheffield: HRI Online Publications, 2014), http://www.hrionline.ac.uk/openbook/chapter/dhc2012-kemman

22 Jeroen Bosman and Bianca Kramer, 'Swiss Army Knives of Scholarly Communication — ResearchGate, Academia, Mendeley and Others', Presentation for STM Innovations Seminar, London, 7 December 2016, https://doi.org/10.6084/m9.figshare.4290428.v1

online affordances of full-text access, hyperlinking between texts, and the subject tags that authors attach.

Moreover, these networks depend on the existing connections between individual scholars. As both authors and readers, academics create online links with one another, becoming followers and followed. This adds a social dimension to the existing dissemination function provided by market-making publishers, by allowing academics to distribute their work via their position in their own disciplinary networks.[23] Besides, or rather countering, the commercial generic platforms, DH scholars increasingly band together in scholarly social networks of their own devising, such as MLA Commons and HCommons.[24] Such close-knit disciplinary connectivity might allow online networks not only to complement the traditional publishers' dissemination services, but outright challenge it. Moreover, through the dissemination of content via social ties between DH scholars, the cohesion within the emergent discipline can be strengthened.[25]

The online environment's inherent properties of a flattened hierarchy and interactive networks also fundamentally affect the function of disseminating texts to different types of audiences. In itself, the notion that authors address specific audiences other than their direct peers is not at all new to the digital medium. Textbooks created for undergraduate students, for instance, are disseminated differently than monographs intended for peer specialists. Such differential targeting simply continues in the distribution of diversified

---

23   That Mendeley is owned by the RELX Group does not subtract from our argument. The publisher does not play a role in the dissemination processes on that platform, although it profits from its functions through data collection.

24   Kathleen Fitzpatrick, 'Academia, Not Edu', *Planned Obsolescence* (26 October 2015), https://kfitz.info/academia-not-edu/; MLA Commons, *An Online Community for MLA Members*, http://mla.commons.org; Humanities Commons, *Open Access, Open Source, Open to All*, http://hcommons.org

25   Cohesive disciplinary networks may help in the effective dissemination of papers, but they also pose the danger of generating more attention for work by eminent scholars (who have many 'followers') than for potentially equally valid work by lesser-known researchers. This Matthew effect (coined as such by Merton in 1968) might threaten adherence to the norm of universality, but since this is a phenomenon not exclusively connected to the functions of formal publication, we will not further engage with it here. See: Robert K. Merton, 'The Matthew Effect in Science', *Science*, 159.3810 (1968), 56–63, https://doi.org/10.1126/science.159.3810.56; James A. Evans, 'Electronic Publication and the Narrowing of Science and Scholarship', *Science*, 321.5887 (2008), 395–99, https://doi.org/10.1126/science.1150473

products through online channels. Yet, besides these existing channels, online platforms have emerged where different interested audiences converge, and communication between them is facilitated. These platforms typically offer a variety of communication types, each with their own rhetoric and degree of complexity: tweets and event announcements appear amidst teaching materials and blog posts on Humanities, Arts, Science, and Technology Alliance and Collaboratory (HASTAC); peer review reports and journal articles feature beside available collaborators and project overviews on DHCommons.[26] This offers the potential to connect with multiple audiences in one environment, and might facilitate cross-dissemination between peers and professionals, including students and interested members of the general public — audiences that humanities scholars aim to address more than other academic disciplines,[27] and that research evaluation frameworks consider increasingly important.[28] The use of broad platforms to disseminate formal publications alongside other types of content intended for other audiences is thus an adaptation of the traditional function of dissemination, again complementary to continuing traditional processes, but with a formative potential for communication practices in DH.

Registration

An extended functionality compared to the print-based tradition can also be observed in the process of registration. Not only has the online medium provided lossless copying at low incremental cost, and low-cost storage and distribution, it has also introduced the technology to accommodate scholarly communication products that were cumbersome or impossible to produce in print. Now, non-textual forms — such as moving images, sounds, or three-dimensional

---

26 HASTAC, https://hastac.org; DHCommons, https://dhcommons.org (link not active at time of publication).
27 Wolff, Rod, and Schonfeld, *UK Survey*, pp. 45–49, esp. fig. 27.
28 For an analysis of research evaluation frameworks' shift towards societal impact, consult: Steven Hill, 'Assessing (for) Impact: Future Assessment of the Societal Impact of Research', *Palgrave Communications*, 2 (2016), https://www.nature.com/articles/palcomms201673; and Teresa Penfield et al., 'Assessment, Evaluations, and Definitions of Research Impact: A Review', *Research Evaluation*, 23.1 (2014), 21–32, https://doi.org/10.1093/reseval/rvt021

objects — can be produced and distributed online in such a way that credit for them can be registered. Semi-textual materials not intended to be read linearly, such as software code and research data, can hardly be made suitable for publication in print, but the affordances of content access and links allow them to be communicated effectively online. Many disciplines, including DH, have witnessed the rise of a rich supply of research products like raw data sets, visualisations, and software, which can now be made available relatively affordably and easily.

The possibility of communicating images, software, and data alongside or as part of formal publications ('enhanced' forms of books and articles) challenges the exclusivity of that formal status resulting from registering authorship that was long reserved for published texts. Now that data, software, and visuals can be made public in their own right, the function of registration, in particular, seems in need of being extended to include 'makership' claims other than authorship in the current legal sense, and ownership claims over objects other than formal publications. Calls for such redefinitions are indeed heard from DH among other disciplines.[29] Besides voicing explicit requests for the reassessment of the notion of authorship, scholars have already begun to extend the definition quite naturally in practice by registering as creators of these new content types and acknowledging authorship of data sets and open source software. Even editable and reusable born-digital content can thus come with authorship claims similar to those of print, without necessarily attaching the same ownership claims as in the print paradigm.

The extension and redefinition of authorship and of the concept of registration of knowledge contributions in any form is thus already

---

29 Harriett Green, Angela Courtney, and Megan Senseney, 'Humanities Collaborations and Research Practices: Investigating New Modes of Collaborative Humanities Scholarship', *Proceedings of the Charleston Library Conference* (2016), https://doi.org/10.5703/1288284316482. For analysis from the digital humanities, see Nowviskie, 'Where Credit Is Due'; Kathleen Fitzpatrick, 'The Digital Future of Authorship: Rethinking Originality', *Culture Machine*, 12 (2011), https://culturemachine.net/wp-content/uploads/2019/01/6-The-Digital-433-889-1-PB.pdf. Similar considerations have been made in other research disciplines in earlier years: Blaise Cronin, 'Hyperauthorship: A Postmodern Perversion or Evidence of a Structural Shift in Scholarly Communication Practices?', *Journal of the Association for Information Science and Technology*, 52.7 (2001), 558–69, https://doi.org/10.1002/asi.1097

taking place, but inevitably finds itself under ongoing assessment and comparison with conventional practice, where authorship registered with formal publications is already (relatively) clearly defined.[30] This is particularly explicit in DH, where there is strong advocacy for attaching value to the registration of work by web-designers, data-analysts, code compilers, and other people who are indispensable in the research process, but who would not be included in the traditional definition of an author.[31] DH scholars especially, more than traditional humanists, find themselves in different roles in the research process: as the principal theorist in their own project, but also beta-testing another's software, or contributing to, enriching or cleaning existing data. Some activities, such as creating an online edition, implicitly assign multiple roles to the scholarly author. The broad digital platforms that allow linking to multiple types of research products (Academia. edu, DHCommons) already facilitate registration in these different roles; and even traditional, print-based publishers are experimenting with mechanisms for acknowledging contributor roles other than authorship.[32] Moreover, registered broad experience and a variety of contributions enhance authors' positions in the social network, which

---

30 We say 'relatively clear', because interpretations of authorship have always varied between the academic fields, as is demonstrated, for instance, by the many different customs for listing co-authorship and for the registration of editors and translators; see for instance: Jenny Fry et al., *Communicating Knowledge: How and Why UK Researchers Publish and Disseminate their Findings*, Research Information Network Report (London: The Research Information Network, 2009), pp. 24–27, http://citeseerx.ist.psu.edu/viewdoc/download?doi=10.1.1.214.8401&rep=rep1&type=pdf

31 Julia Flanders, 'Time, Labor, and "Alternate Careers" in Digital Humanities Knowledge Work', in *Debates in the Digital Humanities*, ed. by Matthew K. Gold (Minneapolis: Minnesota University Press, 2012), pp. 292–308, https://doi.org/10.5749/minnesota/9780816677948.003.0029

32 CRediT, or 'Contributor Roles Taxonomy', is an initiative of the Wellcome Trust, MIT, Digital Science, and several other partners. The taxonomy has been developed with the assistance of CASRAI (Consortia Advancing Standards in Research Administration) and the National Information Standards Organization (NISO), and has to date been implemented in 'badges' that are in use by several publishers, mostly in STEM-fields. The taxonomy itself can be found at http://dictionary.casrai.org/Contributor_Roles; for more information on the CRediT-project and implementations of the taxonomy, see: Liz Allen, 'Moving beyond Authorship: Recognizing the Contributions to Research', *BioMed Central Research in Progress Blog* (28 September 2015), https://blogs.biomedcentral.com/bmcblog/2015/09/28/moving-beyond-authorship-recognizing-contributions-research/; Amye Kenall, 'Putting Credit Back into the Hands of Researchers', *(GIGA)Blog* (28 September 2015), http://gigasciencejournal.com/blog/putting-credit-hands-researchers/

may facilitate dissemination, even if the work itself is not yet valorised in academic evaluation systems.

Although the aim to register all contributors' work is laudable, the intensive involvement of several types of specialists, in itself, is not new in research. Tasks like software compiling or 3D-modelling, at times fulfilled by DH scholars, are inherent in the innovations of the digital medium, but others, such as content design, index creation, and data presentation, resemble services to scholarship that in the print tradition would have been performed by publishing houses, or their freelancers or subcontractors. It should be noted that publishers have already used a function of registration for these services similar to the claims of the author: the publisher brands its products to enhance its reputation by showcasing the excellence of its services. In such instances, the function of registration does not actually change from implicit to explicit, but, as in the case of self-publishing, it shifts from the publisher to the less simply recognisable individual scholar.

One quite fundamental challenge for the function of registration remains: the question of what to register, precisely. Even if the adage that 'scholarship is never finished' was already current in the print age, the submission of a text for publication does, nevertheless, clearly mark the finalisation of a phase or a project. The published version of the text registers its knowledge claims in a finite, stable form. Authors can subsequently add to those claims, challenge them, or refute them in other publications — but the initial registration is not undone. Digital projects on the other hand may develop iteratively and continuously rather than in linear succession of distinct phases. Since online content can be altered or substituted completely following the implementation of newly available insights, online research communication often resembles taking a snapshot of a moving target. This easy adaptation has the advantage of the quick substitution of outdated knowledge — incidentally adding to the perception of increased speed in communication, and perhaps knowledge production itself. At the same time, it fundamentally challenges the function of registration in communication, as it alters the connection between scholars and their individual contributions that had been stabilised in print.[33]

---

33 The adaptable nature of digital objects points towards certain challenges in archiving as well, which will be addressed below in the section on 'Archiving'.

Certification

The two traditional mechanisms for certification are pre-publication review and post-publication citations. Through highly selective filtering and strict quality control, publishers, with the help of academic editors, build their reputation in academia, and authors depend on that reputation to certify their contributions to knowledge.[34] Post-publication certification depends on being cited by peers, departing from the, not uncontested, premise that they will reference high-quality, relevant research only. The digital medium's inherently quantitative nature — the computer is a counting machine after all — has stimulated the use of citation metrics, which is now pervasive in research evaluation. But it has also generated an unprecedented array of complementary instruments of certification for authors. Download counts, page views, shares, likes, bookmarks, retweets, and Wikipedia mentions, to name just a few, offer potential proxies for perceived quality, all equally based on metrics. These 'alternative metrics' have become abundant in social scholarly networks and are increasingly implemented on publishers' platforms.[35]

Like the immediate changes in dissemination and registration, this shift in certification still departs from the existing standard, that is, formal publications. Alternative metrics — such as download counts and link shares — now extend to novel communication forms and even individuals, but have been primarily compiled for books and articles, and they complement rather than substitute existing certification

---

[34] Survey results suggest that humanities scholars value selectivity more than academics from other disciplines, see for instance: Ross Housewright, Roger C. Schonfeld, and Kate Wulfson, *UK Survey of Academics 2012* (Ithaka S+R | Jisc | RLUK, 16 May 2013), pp. 70–72, http://doc.ukdataservice.ac.uk/doc/7644/mrdoc/pdf/7644_uk_survey_of_academics_2012.pdf, https://doi.org/10.18665/sr.22526. Scholars perceive publishers' reputations as important too, but the assumptions on which they build their intuitions remain curiously under-researched. One exploration is made by: Alesia Zuccala et al., 'Can We Rank Scholarly Book Publishers? A Bibliometric Experiment with the Field of History', *Journal of the American Society for Information Science and Technology*, 66.7 (2015), 1333–47, https://doi.org/10.1002/asi.23267. Another suggestion, based on the business operations of book-publishing, is offered by Rick Anderson, *Scholarly Communication: What Everyone Needs to Know* (New York: Oxford University Press, 2018), pp. 181–82.

[35] Although platforms and publishers use in-house technology to compile metrics, many, among which Taylor & Francis, Elsevier, and Oxford University Press, use integrated widgets developed by the enterprise Altmetric, https://altmetric.com

mechanisms. However, considering that the premises on which the traditional proxies of certification rely are themselves contested, increasing use of alternative metrics should be approached at least as critically.[36] They might increase the danger of conflating popularity with authority. Bookmarking or downloading does not equal reading, while reading has never equalled approval, and even citation can indicate violent disagreement. As a direct consequence of the digital medium's salient properties — which cause the Internet's two-way traffic to be logged by default — and publishers' commercial incentives to feature alternative metrics prominently alongside publications, the ample availability of quantitative indicators thus destabilises traditional, much less visible certification.

For digital research results that are disseminated without the involvement of a traditional publisher, further new forms of certification are emerging. Comparable to the brand of the publisher, which signifies authority in print, web projects are stamped with logos of institutional and governmental funders and supporters that are likewise intended to indicate that the communicated research has undergone filtering and quality control. In DH, platforms like NINES and RIDE do not act as publishing venues, but imitate traditional certification by implementing traditional peer review procedures for digital objects aggregated from already existing, but unchecked sources.[37] Also, the uptake of instruments and technology by respected peers may attach value to them, since wide use is regarded as reflecting quality and impact. The DH community boasts many examples, of which the universal acknowledgement of TEI (Text Encoding Initiative) as the de facto standard for text encoding is probably the longest standing.[38]

---

36 Stefanie Haustein, Rodrigo Costas, and Vincent Larivière, 'Characterizing Social Media Metrics of Scholarly Papers: The Effect of Document Properties and Collaboration Patterns', *PLOS ONE*, 10.3 (2015), https://doi.org/10.1371/journal.pone.0120495; James Wilsdon et al., *Next-Generation Metrics: Responsible Metrics and Evaluation for Open Science*, Report of the European Commission Expert Group on Altmetrics (Luxembourg: Publications Office of the European Union, 2017), pp. 12–13, https://ec.europa.eu/research/openscience/pdf/report.pdf

37 *NINES: Networked Infrastructure for Nineteenth-Century Electronic Scholarship*, http://www.nines.org; *RIDE: A Review Journal for Digital Editions and Resources* (IDE), http://ride.i-d-e.de

38 Lou Burnard, 'The Evolution of the Text Encoding Initiative: From Research Project to Research Infrastructure', *Journal of the Text Encoding Initiative*, 5 (2013), https://doi.org/10.4000/jtei.811. At the same time, the TEI also illustrates the registration

Alternative metrics, institutional endorsements through acknowledgement in evaluation systems, and wide uptake are all new forms of traditional types of certification. The importance that clearly attaches to them is evidence that quality control remains crucial in scholarly communication. Such evidence also comes from the rise of a new, fundamentally digital type of certification through networked interactions and iterative versioning. In traditional formal publishing, authors and readers are aware that quality control takes place, but they do not have access to the process: it is a 'black box'. DH is known for its early attempts at opening up this 'black box' of quality control, in one-off experiments such as with *Shakespeare Quarterly* in 2010, or implemented in novel procedures for all publications such as with MediaCommons Press.[39] By providing insight into peers' interactions with texts, open peer review thus, potentially, changes the function of certification: rather than the assertion *that* it has been done, the process of *how* it is done gains importance. These open review procedures still require the optimisation of efforts and gains, as the untimely termination of some experiments perhaps illustrates.[40] Yet, analysis of online engagement with texts is a promising rival to existing certification mechanisms.

Archiving

A stable and dependable apparatus for archiving and retrieving novel communication forms is still lacking. Authors have never really cared greatly about the function of archiving. They have never been actively involved in the process but have traditionally been able to rely on the inherent property of print that multiple copies are distributed widely in a fixed material form, and on the corresponding existing infrastructures, such as library catalogues and publishers' archives. Yet

---

issue discussed earlier. Over time a long list of distinguished but often barely acknowledged scholars have made major contributions to the TEI guidelines.

39 Kathleen Fitzpatrick and Katherine Rowe, 'Keywords for Open Peer Review', *Logos*, 21.3/4 (2010), 133–41, https://doi.org/10.1163/095796511X560024; Media Commons Press, *Open Scholarship in Open Formats*, http://mcpress.media-commons.org/

40 For instance, DHThis, a platform based on a Slashdot-model of user engagement, was launched in 2014, but suspended in 2016 due to lack of interest. Adeline Koh, 'DHThis: An Experiment in Crowdsourcing Review in the Digital Humanities', *Ada: A Journal of Gender, New Media, and Technology*, 4 (2014), https://doi.org/10.7264/N3RX99C5; Bethany Nowviskie argues in 'Where Credit is Due' for a refinement of the processes.

they ought, in their own interest, to take archiving more seriously. As explained above, the online affordances of converging modalities and virtually unlimited storage capacity have expanded authors' potential use of dissemination and registration functions. Besides traditional textual forms, presentations, data, and visuals can now be deposited, for instance, on YouTube or Figshare, and on stand-alone personal or project-based websites. However, this has introduced the problem of digital longevity. If solving this issue on an institutional or national level is already proving a major challenge, how can individual scholars be trusted to solve it satisfactorily? If scholars-as-readers are unsure if they can depend on stable references to such online materials, they may even refrain from citing them altogether.[41] This is not surprising considering that it has taken the traditional infrastructures of scholarly publishing centuries to develop their prized stability and predictability. The limitations of the archival function for novel forms of communication thus pose an immediate and urgent challenge for scholars from disciplines like the digital humanities, who take pride in generating and using them. Fortunately, publishers, libraries, and research funding bodies are increasingly accommodating the archiving of data besides formal publications, as the emerging data archiving policies and principles for fair use demonstrate.[42] These parties seem, from historical contingency, best equipped to generate such archiving functionalities, and authors should be actively involved in advocating their interests.

Conclusions

The digital revolution changes the way knowledge is created, in the humanities as well. The way research results are communicated needs to change accordingly. The reality is that this change happens more slowly

---

41 Although acceptance seems to be growing slightly, researchers report citing far fewer online sources than articles and books. The authority of the cited documents seems to be the main motivation for this. Fry et al., *Communicating Knowledge*, pp. 28–29; see also Fry et al., *Communicating Knowledge* ('Supporting Paper 2: Report of Focus Group Findings', pp. 59–68); Fenlon et al., 'Humanities Scholars', pp. 161–63.

42 One promising example is Force11, an organically grown community of researchers, funders, publishers, and information management professionals that has issued the 'FAIR' principles for research data, which are being increasingly widely adopted. See: Force11.org, *The Fair Data Principles*, https://www.force11.org/group/fairgroup/fairprinciples

than many, especially those in the DH community, would like it to. Over the course of more than four centuries of print, all stakeholders involved in scholarly communication have come to adopt articles and books as the embodiment of all the relevant functions of formal publication: dissemination, registration, certification, and archiving. The established procedures for formal publication; the roles of authors, publishers and libraries in them; and the implicit assumptions about the relationships between these agents have become engrained in the culture of academia to such an extent that we tacitly and automatically rely on the authority of books and articles, instead of weighing the relative importance of each of the functions of publishing in every instance of communication. This can be summarised as the social contract of publishing.[43] Even though scholarly publishing is self-regulating, wherein scholars themselves can change the rules, changing the rules is a matter of patience. There are many partners who are bound by the social contract, and many more fields besides DH.

Even the DH field itself is far from homogeneous. Though there is no hard evidence to prove it, it may well be that, paradoxically, younger DH scholars, for example, experience the stranglehold of this social contract much more acutely than their more senior colleagues.[44] Their career prospects depend on compliance with the existing research evaluation requirements, whereas the communications practices of 'tenured' senior academics are less restricted (although they are not entirely free to do as they please either, as they must maintain the reputations they have built). If so, their conservativeness would act as one more social brake on the adoption of new scholarly communication practices.

Online technologies have expanded the possibilities for scholarly communication. In a much less direct way, they also challenge the existing social acknowledgement of the constellation of functions in formal publications. With so many new communication forms at their disposal, authors are prompted to consider the differences between

---

[43] Peter Drucker theorised the social contract, although not for publishing. See also: Dan Cohen, 'The Social Contract of Scholarly Publishing', *DanCohen.org* (3 March 2010), http://www.dancohen.org/2010/03/05/the-social-contract-of-scholarly-publishing/

[44] Nancy L. Maron, and Sarah Pickle, *Sustaining the Digital Humanities Host Institution Support beyond the Start-Up Phase* (New York: Ithaka S+R, 2014), pp. 15–16, https://doi.org/10.18665/sr.22548; David Nicholas et al., 'So, are Early Career Researchers the Harbingers of Change?', *Learned Publishing*, 32.3 (2019), 237–47, https://doi.org/10.1002/leap.1232

them, and remind themselves, on a more fundamental level, what functions research communication serves in general. In the research fields that study science and scholarship, this draws renewed attention to the values of good scholarship that authors, albeit largely implicitly and unwittingly, uphold by making their research public, regardless of the *form* of publication they choose.

It is precisely the intertwining of the values of scholarship and the functions of traditional publishing, and their ratification in research evaluation systems, that render scholarly communication capable of changing only slowly — even in a field that seems so perfectly suited for quick, disruptive, and radical change as DH. It may be as Kathleen Fitzpatrick has put it, that '[t]he particular contribution of the Digital Humanities [...] lies in the exploration of the difference that the digital can make to the kinds of work that we do, *as well as to the ways in which we communicate to one another*'.[45]

Many of the informal types of communication that used to be entirely private between the instigator and addressee (such as letters, faxes, and telephone calls, but even, for example, conference presentations) are now public by default as a direct consequence of the digital medium's salient properties. But being public does not equate with being published. Some of these new and informal forms of communication might, in due course, become elevated and distinguished with the title 'publication', if they demonstrably serve to uphold values of scholarship — either the traditional Mertonian ones or new ones yet to be established — and if both authors and audiences perceive their function as such. This process of the sociocultural recognition of the online medium's affordances takes time and effort. As we have seen, the DH field's innovative research methods inherently cause it to experiment with new and often initially informal forms of communication, because these serve the functions of publishing as the field intends them to be served. Moreover, the DH community has also shown itself to be good at reflecting on the value of new communication types, along with the necessary reflection on the field's own raison d'être. What remains necessary is building consensus about the value of any new practices that are adopted, and

---

45 Kathleen Fitzpatrick, 'The Humanities, Done Digitally', in *Debates in the Digital Humanities*, ed. by Matthew K. Gold (Minneapolis: University of Minnesota Press, 2012), pp. 12–15 (p. 14, emphasis added), https://doi.org/10.5749/minnesota/9780816677948.003.0002

communicating the result to employers and funders. In the online environment, print-like forms might serve informal communication purposes, while innovative forms might fulfil the same functions as traditional formal publications. To come to an appreciation of good scholarship, in whatever form it may come, we will need a fundamental reconsideration of the traditional, print-based intertwining of form and function of publication. This requires a concerted effort — and time.

Bibliography

Allen, Liz, 'Moving beyond Authorship: Recognizing the Contributions to Research', *BioMed Central Research in Progress Blog* (28 September 2015), https://blogs.biomedcentral.com/bmcblog/2015/09/28/moving-beyond-authorship-recognizing-contributions-research/

Anderson, Rick, *Scholarly Communication: What Everyone Needs to Know* (New York: Oxford University Press, 2018).

Antonijević, Smiljana, and Ellysa Stern Cahoy, 'Researcher as Bricoleur: Contextualizing Humanists' Digital Workflows', *Digital Humanities Quarterly*, 12.3 (2018), http://www.digitalhumanities.org/dhq/vol/12/3/000399/000399.html

Becher, Tony, and Paul Trowler, *Academic Tribes and Territories: Intellectual Enquiry and the Culture of Disciplines* (Buckingham: Society for Research into Higher Education & Open University Press, 2001).

Bosman, Jeroen, and Bianca Kramer, 'Swiss Army Knives of Scholarly Communication — ResearchGate, Academia, Mendeley and Others', Presentation for STM Innovations Seminar, London, 7 December 2016, https://doi.org/10.6084/m9.figshare.4290428.v1

Burnard, Lou, 'The Evolution of the Text Encoding Initiative: From Research Project to Research Infrastructure', *Journal of the Text Encoding Initiative*, 5 (2013), https://doi.org/10.4000/jtei.811

Cohen, Dan, 'The Social Contract of Scholarly Publishing', *DanCohen.org* (3 March 2010), http://www.dancohen.org/2010/03/05/the-social-contract-of-scholarly-publishing/

Cronin, Blaise, 'Hyperauthorship: A Postmodern Perversion or Evidence of a Structural Shift in Scholarly Communication Practices?', *Journal of the Association for Information Science and Technology*, 52.7 (2001), 558–69, https://doi.org/10.1002/asi.1097

Edmond, Jennifer, 'Collaboration and Infrastructure', in *A New Companion to Digital Humanities*, ed. by Susan Schreibman, Ray Siemens, and John

Unsworth (London: Wiley Blackwell, 2015), pp. 54–66, https://doi.org/10.1002/9781118680605.ch4

Evans, James A., 'Electronic Publication and the Narrowing of Science and Scholarship', *Science*, 321.5887 (2008), 395–99, https://doi.org/10.1126/science.1150473

Fenlon, Katrina, et al., 'Humanities Scholars and Library-Based Digital Publishing: New Forms of Publications, New Audiences, New Publishing Roles', *Journal of Scholarly Publishing*, 50.3 (2019), 159–82, https://doi.org/10.3138/jsp.50.3.01

Fitzpatrick, Kathleen, 'Academia, Not Edu', *Planned Obsolescence* (26 October 2015), https://kfitz.info/academia-not-edu/

—— 'The Digital Future of Authorship: Rethinking Originality', *Culture Machine*, 12 (2011), https://culturemachine.net/wp-content/uploads/2019/01/6-The-Digital-433-889-1-PB.pdf

—— 'The Humanities, Done Digitally', in *Debates in the Digital Humanities*, ed. by Matthew K. Gold (Minneapolis: University of Minnesota Press, 2012), pp. 12–15, https://doi.org/10.5749/minnesota/9780816677948.003.0002

Fitzpatrick, Kathleen, and Katherine Rowe, 'Keywords for Open Peer Review', *LOGOS*, 21.3/4 (2010), 133–41, https://doi.org/10.1163/095796511X560024

Flanders, Julia, 'Time, Labor, and "Alternate Careers" in Digital Humanities Knowledge Work', in *Debates in the Digital Humanities*, ed. by Matthew K. Gold (Minneapolis: Minnesota University Press, 2012), pp. 292–308, https://doi.org/10.5749/minnesota/9780816677948.003.0029

Fry, Jenny, et al., *Communicating Knowledge: How and Why UK Researchers Publish and Disseminate their Findings*, Research Information Network Report (London: The Research Information Network, 2009), http://citeseerx.ist.psu.edu/viewdoc/download?doi=10.1.1.214.8401&rep=rep1&type=pdf

Grandjean, Martin, 'A Social Network Analysis of Twitter: Mapping the Digital Humanities Community', *Cogent Arts & Humanities*, 3.1 (2016), 1171458, https://doi.org/10.1080/23311983.2016.1171458

Green, Harriet, Angela Courtney, and Megan Senseney, 'Humanities Collaborations and Research Practices: Investigating New Modes of Collaborative Humanities Scholarship', *Proceedings of the Charleston Library Conference* (2016), https://doi.org/10.5703/1288284316482

Haustein, Stefanie, Rodrigo Costas, and Vincent Larivière, 'Characterizing Social Media Metrics of Scholarly Papers: The Effect of Document Properties and Collaboration Patterns', *PLOS ONE*, 10.3 (2015), https://doi.org/10.1371/journal.pone.0120495

Hill, Steven, 'Assessing (for) Impact: Future Assessment of the Societal Impact of Research', *Palgrave Communications*, 2 (2016), https://www.nature.com/articles/palcomms201673

Holmberg, Kim, and Mike Thelwall, 'Disciplinary Differences in Twitter Scholarly Communication', *Scientometrics*, 101.2 (2014), 1027–42, https://doi.org/10.1007/s11192-014-1229-3

Housewright, Ross, Roger C. Schonfeld, and Kate Wulfson, *UK Survey of Academics 2012*, Ithaka S+R | Jisc | RLUK ([n.p.], 2013), http://doc.ukdataservice.ac.uk/doc/7644/mrdoc/pdf/7644_uk_survey_of_academics_2012.pdf, https://doi.org/10.18665/sr.22526

Hughes, Lorna, Panos Constantopoulos, and Costis Dallas, 'Digital Methods in the Humanities: Understanding and Describing their Use across the Disciplines', in *A New Companion to Digital Humanities*, ed. by Susan Schreibman, Ray Siemens, and John Unsworth (London: Wiley Blackwell, 2015), pp. 150–70, https://doi.org/10.1002/9781118680605.ch11

Kaltenbrunner, Wolfgang, 'Reflexive Inertia: Reinventing Scholarship through Digital Practices' (unpublished doctoral dissertation, Leiden University, Leiden, 2015), https://openaccess.leidenuniv.nl/handle/1887/33061

Kemman, Max, Martijn Kleppe, and Stef Scagliola, 'Just Google It: Digital Research Practices of Humanities Scholars', in *Proceedings of the Digital Humanities Congress 2012*, ed. by Clare Mills, Michael Pidd, and Esther Ward (Sheffield: HRI Online Publications, 2014), arXiv:1309.2434, https://www.hrionline.ac.uk/openbook/chapter/dhc2012-kemman

Kenall, Amye, 'Putting Credit Back into the Hands of Researchers', *(GIGA)Blog* (28 September 2015), http://gigasciencejournal.com/blog/putting-credit-hands-researchers/

Kirschenbaum, Matthew G., 'What is Digital Humanities and What's It Doing in English Departments?' in *Debates in the Digital Humanities*, ed. by Matthew K. Gold (Minneapolis: University of Minnesota Press, 2012), pp. 3–11, https://doi.org/10.5749/minnesota/9780816677948.003.0001

Koh, Adeline, 'DHThis: An Experiment in Crowdsourcing Review in the Digital Humanities', *Ada: A Journal of Gender, New Media, and Technology*, 4 (2014), https://doi.org/10.7264/N3RX99C5

Maron, Nancy L. and Sarah Pickle, *Sustaining the Digital Humanities Host Institution Support beyond the Start-Up Phase* (New York: Ithaka S+R, 2014), https://doi.org/10.18665/sr.22548

Merton, Robert K., *The Sociology of Science: Theoretical and Empirical Investigations*, ed. by Norman W. Storer (Chicago: Chicago University Press, 1973).

—— 'The Matthew Effect in Science', *Science*, 159.3810 (1968), 56–63, https://doi.org/10.1126/science.159.3810.56

Nicholas, David, et al., 'So, are Early Career Researchers the Harbingers of Change?', *Learned Publishing*, 32.3 (2019), 237–47, https://doi.org/10.1002/leap.1232

NINES: Networked Infrastructure for Nineteenth-Century Electronic Scholarship, http://www.nines.org

Nowviskie, Bethany, 'Where Credit Is Due: Preconditions for the Evaluation of Collaborative Digital Scholarship', *Profession* (2011), 169–81, https://doi.org/10.1632/prof.2011.2011.1.169

Nyhan, Julianne, and Oliver Duke Williams, 'Joint and Multi-Authored Publication Patterns in the Digital Humanities', *Literary and Linguistic Computing,* 29.3 (2014), 387–99, https://doi.org/10.1093/llc/fqu018

Ossenblok, Truyken L. B., Frederik T. Verleysen, and Tim C. E. Engels, 'Co-authorship of Journal Articles and Book Chapters in the Social Sciences and Humanities (2000–2010)', *Journal of the American Society for Information Science and Technology,* 65.5 (2014), 882–97, https://doi.org/10.1002/asi.23015, http://onlinelibrary.wiley.com/doi/10.1002/asi.23015/abstract

Penfield, Teresa, et al., 'Assessment, Evaluations, and Definitions of Research Impact: A Review', *Research Evaluation,* 23.1 (2014), 21–32, https://doi.org/10.1093/reseval/rvt021

Praal, Fleur, and Adriaan van der Weel, 'Taming the Digital Wilds: How to Find Authority in a Digital Publication Paradigm', *TXT,* 4 (2016), 97–102, https://openaccess.leidenuniv.nl/bitstream/handle/1887/42724/PraalvdWeel.pdf

Prosser, David C., 'Researchers and Scholarly Communications: An Evolving Interdependency', in *The Future of Scholarly Communication,* ed. by Deborah Shorley and Michael Jubb (London: Facet, 2013), pp. 39–49, https://doi.org/10.29085/9781856049610.005

Purdy, James P., and Joyce R. Walker, 'Valuing Digital Scholarship: Exploring the Changing Realities of Intellectual Work', *Profession* (2010), 177–95, https://doi.org/10.1632/prof.2010.2010.1.177

Quan-Haase, Anabel, Kim Martin, and Lori McCay-Peet, 'Networks of Digital Humanities Scholars: The Informational and Social uses and Gratifications of Twitter', *Big Data & Society* (2015), pp. 1–12, https://doi.org/10.1177/2053951715589417

RIDE: A Review Journal for Digital Editions and Resources (IDE), http://ride.i-d-e.de

Roosendaal, H. E., and P. A. Th. M. Geurts, 'Forces and Functions in Scientific Communication: An analysis of their interplay', unpublished conference paper at Cooperative Research Information Systems in Physics, Oldenburg, Germany, 31 August–4 September 1997, http://www.physik.uni-oldenburg.de/conferences/crisp97/roosendaal.html

Spiro, Lisa, '"This is Why We Fight": Defining the Values of the Digital Humanities', in *Debates in the Digital Humanities,* ed. by Matthew K. Gold (Minneapolis: University of Minnesota Press, 2012), pp. 16–36, https://doi.org/10.5749/minnesota/9780816677948.003.0003

Weel, Adriaan van der, *Changing Our Textual Minds: Towards a Digital Order of Knowledge* (Manchester: Manchester University Press, 2011).

Weel, Adriaan van der, and Colleen Campbell, 'Perspectives on the Future of the Monograph', *Against the Grain*, 28.3 (June 2016), 1, 10, http://www.against-the-grain.com/wp-content/uploads/2016/07/ATG_v28-3.pdf

Whitley, Richard, *The Intellectual and Social Organization of the Sciences* (Oxford: Oxford University Press, 1989).

Wilsdon, James, et al., *Next-generation Metrics: Responsible Metrics and Evaluation for Open Science*, Report of the European Commission Expert Group on Altmetrics (Luxembourg: Publications Office of the European Union, 2017), https://ec.europa.eu/research/openscience/pdf/report.pdf

Wolff, Christine, Alisa B. Rod, and Roger C. Schonfeld, *UK Survey of Academics 2015*, Ithaka S+R | Jisc | RLUK ([n.p.], 2016), https://doi.org/10.18665/sr.282736

Ziman, John, *Real Science: What It Is and What It Means* (Cambridge: Cambridge University Press, 2000).

Zuccala, Alesia, et al., 'Can we Rank Scholarly Book Publishers? A Bibliometric Experiment with the Field of History', *Journal of the American Society for Information Science and Technology*, 66.7 (2015), 1333–47, https://doi.org/10.1002/asi.23267

3. Academic Publishing

New Opportunities for the Culture of Supply and the Nature of Demand

Jennifer Edmond and Laurent Romary

Introduction

The scholarly monograph has been compared to the Hapsburg monarchy in that it seems to have been in decline forever![1]

It was in 2002 that Stephen Greenblatt, in his role as president of the US Modern Language Association, urged his membership to recognise what he called a 'crisis in scholarly publication'. It is easy to forget now that this crisis, as he then saw it, had nothing to do with the rise of digital technologies, e-publishing, or open access. Indeed, it puts his words into an instructive context to recall that it was only later in that same year that the Firefox browser saw its initial release. The total number of websites available in the world in that year was only around three million, compared to the nearly two billion available today.[2]

What Greenblatt was actually concerned about was the precarious economic viability of the scholarly monograph, and the resulting decline in monograph production by traditional presses, combined with an increasing demand for such monographs from individuals

---

[1] Colin Steele, 'Scholarly Monograph Publishing in the 21st Century: The Future More Than Ever Should Be an Open Book', *Journal of Electronic Publishing*, 11.2 (2008), https://doi.org/10.3998/3336451.0011.201

[2] 'Total Number of Websites: Internet Live Stats', *Internet Live Stats*, http://www.internetlivestats.com/total-number-of-websites/

and institutional hiring, tenure, and promotion committees as a mark of scholarly achievement.[3] Over a dozen years later, not only have the original problems Greenblatt identified not gone away, but a whole raft of further complications have — for scholars, for publishers, and for libraries — also emerged to join them.

Given the long history of this debate, its current focus on the 'digital turn' in scholarly communication perhaps obscures an additional potential area of focus on what one might call the 'supply side' of the equation. The practices we use to produce, release, and otherwise share scholarship are, of course, of great concern and importance to the system of knowledge circulation. Recent work, like that of the *The Academic Book of the Future* project,[4] along with others described in the *Journal of Scholarly Publishing*'s 'Special Issue on Digital Publishing for the Humanities and Social Sciences',[5] have illustrated the breadth of systemic change as well as the multiple players involved and affected by it. Such contributions not only highlight the richness of the emerging landscape of knowledge production, but also the many perspectives that contribute to it, including, but by no means limited to, that of the researcher him or herself. But for all of the plurality and depth these innovative discussions bring to our understanding of how scholarship comes to be produced and made available for further use, what happens to this work afterward remains largely taken for granted. Changes in scholarly communication need to be understood as a two-way process, of both production and consumption; but the latter aspect seems to attract far less attention than the former. Paying unequal attention to this aspect of the overall circulation of knowledge raises the risk of perpetuating traditions of communication practice that may not suit the equally transformed set of information retrieval, reuse, interrogation, and application practices. Form, as John

---

[3] Stephen Greenblatt, 'A Special Letter from Stephen Greenblatt', *Modern Language Association* (28 May 2002), https://www.mla.org/Resources/Research/Surveys-Reports-and-Other-Documents/Publishing-and-Scholarship/Call-for-Action-on-Problems-in-Scholarly-Book-Publishing/A-Special-Letter-from-Stephen-Greenblatt

[4] *The Academic Book of the Future*, ed. by Rebecca Lyons and Samantha Rayner (Basingstoke: Palgrave Macmillan, 2016), https://doi.org/10.1057/9781137595775

[5] 'Special Issue on Digital Publishing for the Humanities and Social Sciences', ed. by Alex Holzman and Robert Brown, *Journal of Scholarly Publishing*, 48.2 (2017), https://doi.org/10.3138/jsp.48.2.73.

Naughton reminds us, should not be conflated with function,[6] and the fact that a certain form has met the requirements of scholars in the past does not mean that a virtual reincarnation of that form will do the same for the scholars of the future.

The 'demand side' of the scholarly communication equation must therefore address the changes in the reading habits of consumers of published research. Before we can do so, however, it is worth pausing briefly to consider exactly what it means, in the current day and age, to 'publish' research, rather than to 'communicate' or 'disseminate' it. Although it may seem that any of these could be used to refer to the key process implied in the etymology of the term (that is, to make something public), publishing is generally agreed to be the most restricted of the terms (though there is much overlap in their general use). According to Leah Halliday, 'scholarly publishing is a means of communicating scholarship within a community';[7] a not very helpful definition in itself, but one that she uses to help tease out the issues of how a work is distributed, its formality, its durability, and in particular its status as validated by the community. These factors lend a particular act of communication (which usually also adheres to certain norms of format and structure, for example, as a monograph or as a journal article) an authority that more informal acts of communication will struggle to establish, but they also imply a set of power relationships that both authors and readers participate in.

Researchers who are seeking to expand their knowledge are, first and foremost, regarded as comprising the cohort of consumers of scholarship; but only slightly upstream from them are the evaluation and assessment panels controlling how the research may be perceived (as, for example, through publication in a well-regarded journal), or how it may be transformed into capital to access rewards at either the institutional level (as in promotion) or externally (as in funding grants). The goals and needs of these two groups — one seeking knowledge the other seeking to assign value — do not necessarily align with each other however, a fact that has been a source of tension since the

---

6   John Naughton, 'The Future of News (and of Lots More Besides)', *Memex 1.1* (17 March 2009), http://memex.naughtons.org/archives/2009/03/17/6998

7   Leah Halliday, 'Scholarly Communication, Scholarly Publication and the Status of Emerging Formats', *Information Research*, 6.4 (2001), http://www.informationr.net/ir/6-4/paper111.html

'original' publication crisis. This issue bridges the divide between the nomadic nature of knowledge creation in the humanities and the academic rewards system to which a given producer of scholarship is bound, taking on as a part of this relationship not just the possibilities for reward, but any perverse incentives it may create.[8]

The Place of the Book in Humanities Communication

This complex relationship between the consumption and production of scholarship can perhaps be nowhere more easily seen than in the status of the book as a specific and privileged instrument for scholarly communication in the humanities. From the perspective of the writer, the reasons for the tenacity of the book are many, and encompass not only the epistemic and intellectual benefits the form provides, but also the more emotional aspects of attachment to the long monograph, to the expansiveness of the prose, the physicality of the book-as-object, and the tangible representation of one's intellectual achievement. Human Computer Interaction (HCI) researchers have explored some of the underlying psychology of this, finding that digital objects are perceived as having a lower value because they are less distinctive and more easily replicated at little cost, as opposed to 'a seashell or crayon drawing [which] is unique in its singular presence'.[9] One can, of course, also find justification for this preference in anthropological work that demonstrates a physical object's ability to embody the owner's identity and personal history.[10]

The drivers behind our attachment to the physical book can be found in the physiological as well as the symbolic. In spite of the continued improvement of computer screens, paper remains a far better carrier of information, holding up to fifty times more information for a given area. Paper also does not suffer from the 'flicker effect', which causes

---

8 Dennis Leech, 'Perverse Incentives Mean the REF Encourages Mediocrity rather than Excellence', *REF Watch* (10 December 2013), http://ref.web.ucu.org.uk/2013/12/10/perverse-incentives-mean-the-ref-encourages-mediocrity-rather-than-excellence/

9 Melanie Feinberg, 'Beyond Digital and Physical Objects: The Intellectual Work as a Concept of Interest for HCI', in *Proceedings of the SIGCHI Conference on Human Factors in Computing Systems* (Paris, France, April 27-May 02, 2013), pp. 3317–26, https://doi.org/10.1145/2470654.2466453

10 Daniel Miller, *The Comfort of Things* (Malden, MA: Polity, 2008) is a good example of where this appreciation of objects can lead.

us to lose up to forty percent of the information presented to us on a computer screen.[11] In the wider reading market, all of these factors can be seen as contributing to the continued strength of the printed versions of books as opposed to those same books in electronic formats, with the eBook market seeming to plateau at twenty-five to thirty percent of total sales.[12] Regardless of the wider trends and the reasons why this is the case, as long as humanities disciplines view themselves, and are viewed by others, as having a 'soul [that] lies between the covers of a scholarly monograph',[13] then the prestige in printed books will remain in the perceived exclusivity of the long form and in its physicality, which the age of e-publishing has yet to effectively supplant.

Recent work on the specific, ideal shape and form of the scholarly book in the digital age has extended our understanding of the unique place it occupies, although the explanations are neither conclusive nor complete. Of particular importance is the 2012 OAPEN survey[14] and the analysis of its results, which appeared in the 2015 report *Monographs and Open Access*.[15] In this latter work, the author, Geoffrey Crossick, lays out an excellent case for the reasons why humanists need to write books, which is largely because the ability to create a sustained discourse is formative for good arguments in the humanities disciplines. This argument in favour of the writing of books is not only true for the humanities; interestingly the same basic argument was put forward in a 2010 *Nature* editorial entitled 'Back to Books', but this time addressing the benefits the writing of books could bring to the hard sciences'

---

11   Edward J. Valauskas, 'Waiting for Thomas Kuhn: First Monday and the Evolution of Electronic Journals', *First Monday*, 2.12 (1997), http://firstmonday.org/ojs/index.php/fm/article/view/567

12   Frank Catalano, 'Paper is Back: Why "Real" Books Are on the Rebound', *GeekWire* (18 January 2015), http://www.geekwire.com/2015/paper-back-real-books-rebound/; Jim Milliot, 'For Books, Print Is Back', *PublishersWeekly.com* (2 January 2015), https://www.publishersweekly.com/pw/by-topic/industry-news/bookselling/article/65172-print-is-back.html

13   Jennifer Wolfe Thompson, 'The Death of the Scholarly Monograph in the Humanities? Citation Patterns in Literary Scholarship', *Libri*, 52.3 (2002), 121–36 (p. 122), https://doi.org/10.1515/LIBR.2002.121.

14   'Survey of Use of Monographs by Academics — as Authors and Readers', *OAPEN-UK* (2014), http://oapen-uk.jiscebooks.org/files/2012/02/OAPEN-UK-researcher-survey-final.pdf. The acronym OAPEN stands for: Open Access Publishing in European Networks.

15   Geoffrey Crossick, *Monographs and Open Access: A Report to HEFCE* (London: HEFCE, 2015), https://dera.ioe.ac.uk/21921/1/2014_monographs.pdf

community.¹⁶ The benefit of writing books seems clear, but the evidence given for Crossick and others' arguments for reuse by readers of this long form is less conclusive: 'nearly two thirds of those responding [to the OAPEN survey] had used a scholarly book for work purposes within the previous week. [...] While only a third of respondents reported that they had read the whole book, only 11 per cent of those surveyed had read one chapter or less'.¹⁷

Crossick views these numbers optimistically as indicators that readers still engage with books as sustained arguments, not as the sum of a set of disassociated parts. Not everyone shares his optimism. OAPEN's 2010 analysis of users' needs relating to digital monographs in the humanities and social sciences presents a somewhat different (and perhaps more cynical) view: 'People do not read books anymore, they read a chapter or a paragraph [...] to read a book from beginning to end is out of fashion. Since you're under pressure to do research, to publish and so on, you don't have time to read anymore. Read or rot doesn't exist, publish or perish does'.¹⁸ Whether this has always been the primary mode for reading scholarly books is, the authors state, unclear, but certainly the affordances and habits of the digital do not militate against such a paradigm for reading while clearly facilitating it in many ways.

Against this backdrop, and in a system and culture where so much of the prestige and traditional shorthand for important work is still tied up with our positive perceptions of traditional books (printed, or digital simulacra of printed), we have to assume that many authors produce books not because this is necessarily the best form of communication for their work, but because they feel this will bring the most benefits, because it is what they have been trained to do, or indeed because they feel they have little choice. Many such externalities seem to influence this choice, as Tim C. E. Engels et al. have shown in terms of funder mandates.¹⁹ The experience of the authors of the *London Lives* study

---

16 'Back to Books', *Nature*, 463 (2010), 588–88, https://doi.org/10.1038/463588a
17 Crossick, *Monographs and Open Access*, p. 22.
18 Janneke Adema and Paul Rutten, *Digital Monographs in the Humanities and Social Sciences: Report on User Needs*, OAPEN Deliverable, 3.1.5 (2010), https://openreflections.files.wordpress.com/2008/10/d315-user-needs-report.pdf, p. 62.
19 Tim C. E. Engels et al., 'Are Book Publications Disappearing from Scholarly Communication in the Social Sciences and Humanities?', *Aslib Journal of Information Management*, 70.6 (2018), 592–607, https://doi.org/10.1108/AJIM-05-2018-0127

in trying to produce an eBook version of their monograph illustrates how this social force can manifest itself as the authors' desire to publish something different becomes diluted by a lack of imagination on the part of the publisher:

> [the publisher's] idea of an eBook was little more than a photographic edition of the printed text. Like most current eBooks, it would essentially have the appearance of a pdf file, with a limited number of external links to trusted sources. And their production methods prioritised the printed book, with the eBook expected to follow obediently behind.[20]

Books may be many things, but the fact that a scholar wanting to present his results in an imaginative format is unable to escape the gravity of the proxies and symbolic capital of the traditional book throws open the question of what we, as readers, really need and want books for.

Scholarly Reading and Browsing

Given our physiological and social attraction to books, the benefits their creation brings in terms of developing the key skills required for scholarship, and the production biases in the system, the evidence that two thirds of books are not consumed in their entirety seems to bear out rather than disprove that the form may no longer be fit for all of the functions it is used for. Authors may bristle at the idea that their publishers are actually willing to sell access to only the introduction and first chapter of their well-crafted monographs; but those same authors are also very likely to consume, as researchers, the work of their peers in exactly the same manner: piecemeal, and only following a path and intensity that suits their own research questions and practices rather than seeking to match that of the author of the work. As John Guillory put the case in his touchstone article on scholars' information consumption practices (and again, even before the digital made such practices so much easier): 'Scholarly books are pulled apart like the Sunday paper'.[21] A book is more than an object, it also represents a mode of communication — a format suited to a

---

20 Bob Shoemaker, 'The Future of the (e)Book', *History Matters* (1 December 2015), http://www.historymatters.group.shef.ac.uk/future-ebook/
21 John Guillory, 'How Scholars Read', *ADE Bulletin*, 146 (2008), 8–17 (p. 14), https://doi.org/10.1632/ade.146.8

complex, contextualised, densely evidenced argument. While it is clear we still deeply respect and inherently require this mode of scholarly expression, it is no longer clear that this is our primary mode for consuming scholarship, nor that it will continue to enjoy the primacy that it does.

There is relatively little evidence to support or refute this claim beyond the OAPEN studies of book readership discussed above. Although it is more focused on researchers' perceptions of the mode of production (of their own work) than of consumption (of the work of others), a 2012 Jisc study showed that when researchers print out electronic resources (primarily book chapters and journal articles, one assumes), they are more likely to do so in part than as a whole.[22] However, even here, we lack evidence for what defines a 'part', or indeed for what is then done with the work that has been printed off, or whether the incentives for printing are directly related to an intention to read, or driven by resource considerations. The challenge of understanding the interaction, or indeed the disconnect, between the needs and choices of the writer of published scholarship and those of the reader remains, despite the fact that most of the people who play one of those roles in the system also plays the other.

Some relevant research on the general behaviours exhibited by users of virtual libraries does exist,[23] and from this body of work two trends in particular emerge that can be viewed as pertinent for the digital age. The first of these is 'horizontal information seeking', which refers to the habit of looking at only a small percentage of a site's content, then navigating away from it (often not to return again). This behaviour seems to be the norm, not the exception. A CIBER/UCL study found that around sixty percent of e-journal users viewed no more than three pages of the journal, and the majority never returned to that source afterward. The second potentially relevant information-gathering trend is 'squirrelling behaviour', which refers to the habit of amassing a significant amount

---

22 Caren Millen, 'Exploring Open Access to Save Monographs, the Question Is — How?', *Jisc*, https://www.jisc.ac.uk/blog/exploring-open-access-to-save-monographs-the-question-is-how-24-oct-2012

23 Ian Rowlands et al., 'Information Behaviour of the Researcher of the Future', *CIBER Briefing Papers* (2008), https://www2.warwick.ac.uk/study/cll/courses/professionaldevelopment/wmcett/researchprojects/dialogue/the_google_generation.pdf

of downloaded material and saving it for later digestion (or not). These are not merely the habits of the younger, 'Google generation', either, as the same study also found: 'from undergraduates to professors, people exhibit a strong tendency towards shallow, horizontal, "flicking" behaviour in digital libraries. Power browsing and viewing appear to be the norm for all'.[24] Needless to say, not all reading behaviours overlap with these more superficial information-seeking strategies, but the likelihood of overlap cannot be ignored.

In some ways, this move from focused consumption to selective browsing seems a natural reaction to the information age. One can imagine that there would have been a time when only privileged access to a great library could have brought a scholar into contact with this many volumes. In this context, a scholarly work of breadth would have represented the consolidation of a field of knowledge, and be of great potential service to readers who might not have the same access to previous work. But the all-encompassing and complete nature that a humanist's knowledge is expected to somehow represent has become enshrined in our modes not just of publishing, but of conceiving our disciplines and our epistemologies. A work of humanistic scholarship is still expected to report a research finding while also deeply contextualising that finding: in essence, it is expected to curate a body of knowledge. This requirement is not based on tradition alone, but on the manner in which humanistic knowledge is created not by experimentation (which is then presented in written form) but, as many argue, in the act of writing itself: 'In the humanities, scholars have tended to be physically alone when at work because their primary epistemic activity is the writing, which by nature tends to be a solitary activity'.[25] But information curation as enacted in these epistemological acts of writing has been disintermediated in the information age, hence the widening gap between our informational behaviours as horizontal browsers, and our attachment to the traditional forms of scholarly communication.

---

24 Ian Rowlands, 'Information Behaviour', p. 19.
25 Willard McCarty, *Humanities Computing* (Basingstoke: Palgrave Macmillan, 2005), p. 12.

Old and New Ways to Share Knowledge

Whether or not the traditional modes of scholarly production and communication that are currently being reproduced to operate in virtual environments, are outdated is one question — whether or not they have any negative effects on scholarship is quite another. Although he does not directly note any disconnect between writers and readers, Clay Shirky points out the irrelevance of past forms of publication for the future:

> With the old economics destroyed, organizational forms perfected for industrial production have to be replaced with structures optimized for digital data. It makes increasingly less sense even to talk about a publishing industry, because the core problem publishing solves — the incredible difficulty, complexity, and expense of making something available to the public — has stopped being a problem.[26]

Shirky's paradigm primarily applies to scholarship to the extent that the optimal unit of communication for scholars is perhaps shifting, and the potential for disaggregating processes formerly seen as interlinked (such as editing, peer review, and distribution) has grown. This is true for books but also for articles, and indeed beyond these, as long-standing, verified forms begin to become peaks in an overall scholarly production that has a very long tail. One of the differentiating aspects, introduced above, between publication and other forms of scholarly communication is that of formality versus informality; but formality is a standard based upon conservative norms and it is inclined to shift from one generation to the next. 'The new way of digital scholarship [is] actively sharing thinking, images, films, etc. to provide primary resources for others'.[27] This idea of 'actively sharing' is not necessarily compatible with the certification and production practices of traditional publication, so the informal communications channels multiply and grow in profile.

---

26   Clay Shirky, 'Newspapers and Thinking the Unthinkable', *Clay Shirky* (13 March 2009), http://www.shirky.com/weblog/2009/03/newspapers-and-thinking-the-unthinkable/

27   Joan Cheverie, Jennifer Boettcher, and John Buschman, 'Digital Scholarship in the University Tenure and Promotion Process: A Report on the Sixth Scholarly Communication Symposium at Georgetown University Library', *Journal of Scholarly Publishing*, 40.3 (2009), 219–30 (p. 225), https://doi.org/10.3138/jsp.40.3.219.

How does this proliferation of forms, which also includes shorter textual communication such as blog posts and tweets, aspire to the equivalent of wearing a suit and tie to satisfy not only the reading audience but the norms of quality control? Some of this hybrid sharing is compatible with the sustained argument form, and the delivery and validation norms of the monograph; but in other cases it might be better produced as a digital edition, exhibition, or performance; as a blog or other form of open or closed, full-length or micro-length publication; a collection of curated and/or annotated links or references; a methodological or teaching resource; a dataset or visualisation; or indeed software, tools, and platforms. But, as Robert Brink Shoemaker's experience of creating the eBook for *London Lives* seems to illustrate, form may not be allowed to follow function, or, at least, it may not be valued by some readers (e.g. evaluators) in the same way as others (e.g. scholars seeking insight). Stated another way, the challenge that faces us is not to do away with long or traditional forms of scholarship, but to supplement them by coming to understand how smaller or different units of scholarly production can accrete to create a sustained argument, or speak with a different language yet still be verifiable; and how the depth of the book can be replicated in some cases and for some topics without simply mimicking, or otherwise creating in another guise, the known form of the monograph.

In part, these new forms challenge our ability to share knowledge: merely making scholarly output available online brings no guarantee that it will find its specialist audience. Perhaps more critical, however, is the difficulty the wider research ecosystem has with validating such scholarship: 'humanities have little excuse for holding on to archaic forms of evaluation that hold back new forms of scholarship because we lack a roadmap for how to attribute credit for work in digital humanities.'[28] Certainly the dependence on the publication of monographs as a marker for scholarly maturity is still harmful in the way that Greenblatt highlighted more than a decade ago. And yet, pillars of the system cling to the primacy of print. For example, in a controversial policy statement, the American Historical Association (AHA) advocated placing a six-year embargo on making PhD theses

---

28   Cheverie, Boettcher, and Buschman, 'Digital Scholarship', p. 226.

digitally available with the following justification: 'History has been and remains a book-based discipline'.[29]

This statement may or may not tell the whole story; as another historian has commented: 'historians tend to be notoriously covetous about whatever they're doing and they don't want to share even within a collaborative context'.[30] This impulse could also be at play here. A further analysis points the finger at a complicity with publishers who are fighting a rear-guard action to defend their business models, leading to what A. Truschke calls: 'this bizarre idea of the unpublished but broadly accessible dissertation'.[31] For whatever reasons, however, and, of course, with some exceptions, the AHA's statement on the place of the book seems generally to be all too true, not just for history but for all of the humanities, including, somewhat ironically, the digital humanities. In a 2011 study carried out by the Research Information Network (RIN), a series of six case studies were presented, each profiling work that had a strong digital component. And yet, repeatedly, when discussing the dissemination practices of the scholars in question, the section on dissemination echoed the same incantation: 'All the respondents in this case disseminate their research primarily through traditional means such as peer-reviewed journals, monographs, chapters in edited books, and conference presentations'.[32] When encouraged to reflect further, each cohort revealed an awareness of other alternatives, and even, at times, an eagerness to avail themselves of them; but there were barriers as well, which ranged from the feeling that the research was not suited to a broad public audience, through to a strong sense that one had to pick one's venues for publication carefully (and conservatively) for career advancement.

---

29 Jacqueline Jones, 'AHA Statement on Policies Regarding the Embargoing of Completed History PhD Dissertations', *American Historical Association* (22 July 2013), http://blog.historians.org/2013/07/american-historical-association-statement-on-policies-regarding-the-embargoing-of-completed-history-phd-dissertations/

30 Lorraine Estelle, 'What Researchers Told Us about their Experiences and Expectations of Scholarly Communications Ecosystems', *Insights*, 30.1 (2017), 71–75, https://doi.org/10.1629/uksg.349

31 A. Truschke, 'Dissertation Embargoes and Publishing Fears', *Dissertation Reviews* (1 April 2015), http://dissertationreviews.org/archives/11842

32 Monica Bulger et al., *Reinventing Research? Information Practices in the Humanities*, Research Information Network Report (London: The Research Information Network, 2011), p. 26.

The following examples provide further evidence for this phenomenon. One faculty member said: 'I still need to improve my publications record. I think once I've managed to get a couple of things in traditional journals I will probably try to move towards the commercial free, open, Internet journals.' Yet many referred to blogs in their descriptions of useful resources. One predicted that in the future blogging would be more acceptable: 'The barrier between real publications as we used to understand them and mere documents on the web is beginning to dissolve [...] or is becoming more permeable.'[33] A generational shift seems to be happening, but largely in parallel with the already longstanding formal, verified, and rewarded communication flows. This pattern continues to be perpetuated despite numerous attempts to publish guidelines, such as the 2011 edition of the MLA's journal *Profession*, with its suite of articles on the evaluation of digital scholarship; and the draft 'Guidelines for the Professional Evaluation of Digital Scholarship in History',[34] published in April 2015. It is possible that, in some ways, this conservatism serves scholarship well, and by maintaining its strongest valorisation for the sustained argument, the systems of evaluation and reward may indeed be maintaining a high standard for the depth and formality of argument. However, it is also possible that this conservatism largely serves another purpose. As one scholar quite pointedly states: 'I think that over the last thirty years literature departments learned how to outsource a key component of the tenure granting process to university presses';[35] and it is these smaller presses that have been hardest hit by the digital transformation of their industry. With this statement, however, we also return to the question of how publication cultures are shaped by the readers and consumers of scholarship who are not seeking knowledge for their own use, but as input for the validation of others.

---

33 Bulger et al., *Reinventing Research*, p. 45.
34 Seth Denbo, 'Draft Guidelines on the Evaluation of Digital Scholarship', *American Historical Association* (21 April 2015), http://blog.historians.org/2015/04/draft-guidelines-evaluation-digital-scholarship/
35 Lindsay Waters, 'A Modest Proposal for Preventing the Books of the Members of the MLA from Being a Burden to their Authors, Publishers, or Audiences', *PMLA*, 115.3 (2000), 315–17 (p. 316), https://doi.org/10.2307/463452.

The Evaluator as an Audience for Scholarship

The different needs for scholarship of an audience that is comprised more of managers of scientific staff and/or research budgets (in reality the same individual may wear different hats) than of researchers come to the fore in this context. Even in a perfect system there could be no perfect evaluation, unless that evaluation could somehow be decoupled from competition and rewards. In addition, such competitions, be they for a slot in a prestigious journal, grant funding, or for a promotion in a system with quotas in place (tacit or explicit), often require those panels who are charged with making the decisions to take into consideration a large amount of information about disparate projects or individuals. It also requires those evaluators to exert great care in managing their potential biases and knowledge gaps in the formulation of their conclusions about what they read, as these conclusions will have a direct impact on a colleague. There is a great temptation to rely on heuristics, proxies, or externalised systems, whether they are impact factors or publisher reputations, to ease and align these processes. Such systems can be robust, albeit usually only within narrow parameters. But they are also open to abuse: the creators of metrics-based approaches and scientific databases very often decry the uses to which they are put,[36] and the fact that the big scientific information databases, Scopus and Thomson ISI, have not readily included monographs has not helped in this respect; although alternative approaches enabling a more balanced metrics-based approach to the humanities have been developed in many countries.[37] In addition, the question of what science should be evaluated for has become more pressing in recent years. Public pressure to deliver value for money has focused the evaluators' attention on the impact (social, industrial, educational) of a piece of research as well as its perceived excellence. Again, measures have been developed to try and quantify this abstract notion, but the rise of another complicating

---

36 Ferenc Kiefer, 'ERIH's Role in the Evaluation of Research Achievements in the Humanities', in *New Publication Cultures in the Humanities Exploring the Paradigm Shift*, ed. by Péter Dávidházi (Amsterdam: Amsterdam University Press, 2014), pp. 173–82, http://www.oapen.org/search?identifier=515678

37 Elea Giminez-Toledo et al., 'The Evaluation of Scholarly Books as Research Output: Current Developments in Europe', in *Proceedings of the 15th International Society for Scientometrics and Informetrics Conference*, 29 June–4 July 2015, Istanbul, Turkey, http://curis.ku.dk/ws/files/141056396/Giminez_Toledo_etal.pdf

factor in the system of scientific measurement (regardless of its potential to align with new forms of scholarly communication) has exacerbated the notion that such evaluations have to defend their methods even as they are coming under increased pressure.

The institutional and cultural barriers are therefore high, and grounded in the very traditions that make humanities research what it is. Maintaining the model of the lone scholar and the long monograph as primary touchstones for scholarly production also raises significant barriers to meeting some of the emergent expectations for scholarship: the need to engage wider audiences, to create clear and auditable trails of scholarship through analogue and digital resources, to meet the moral and financial demands of the emergent open access publishing system, and to be able to recognise quality scholarship as knowledge creation and communication in itself rather than via its proxies such as a book published by a certain publisher. These humanities' quality marks do, in some cases, have a better basis for their role as proxies for quality than some science equivalents, such as the now widely discredited journal impact factors. But their dominance, justified or not, within the minds of disciplinary communities stymies innovation and prevents the optimisation and customisation of the research communication workflow. The drive toward diversity is not an external pressure being applied to humanists, but the will and desire of the research community itself: the outcry criticising the AHA's proposal for a digital thesis embargo was equally as passionate as the original pronouncement. While it may be that the digital transformation is framing the conversation, scholarship is not changing just because of the digital, but rather due to the changing needs and wants of the scholars themselves as readers and users of scholarship, as well as in their role as its producers.

Barriers to Change, and Opportunities

The driving principle behind scholarly communication generally (and publication in particular) should be to maximise the reach and resonance of research results. Carrying out a research activity is all about exploring diverse territories, where knowing what others are doing, what their most recent advances are, and what projects are being undertaken, is essential to making sure that one's own research actually goes beyond the state of the art and can be situated within the larger landscape of insight and discovery. Communicating results is an essential activity in academic life, and not only because the assessment of such communication, through peer review mechanisms, impacts on institutional and funder recognition, thereby facilitating the financial means to carry out further research. However, scholarship is also based upon community consensus, and for this reason the validation functions of publication remain highly relevant.

So, what are the barriers to the widespread uptake of new publishing models that can accelerate the process of scholarship and the sharing of knowledge? In addition to those things discussed above, there are two primary forces that will need to be addressed: protection and authority. The first of these issues is fundamental to the publication and reward system, with the imperative it puts forward that scholars must produce research that is original. In the case of historians, for instance, we can see that they are very protective of their data and their sources, as was found by Diane Harley et al.,[38] until such time as they have published their work — and rightly so, given the close linking of originality of research with reputation, publication, and, by direct extension, tenure and promotion. But while the more rapid communication cycle of the science disciplines may, in many ways, be driven by technological change, it is also underpinned by a system for protecting discovery: through patenting, licensing, and other such instruments. It should, theoretically, be equally possible for an historian to discover links between sources, or uncover unknown sources, and to similarly protect and share this discovery. This kind of research output need only be able to provide a traceable link to the author of the idea and his or her evidence base, something that could be included as a reference by other scholars seeking to build upon this work. If such conventions were in place, there would be no reason why a work of any length could not be considered as an independent 'act of scholarship'.

There is no technical barrier to the rise of such formats: indeed, many platforms and standards for them exist already. Additionally, such micro-publications need never become the whole of a scholar's output,

---

38 Diane Harley et al., *Assessing the Future Landscape of Scholarly Communication: An Exploration of Faculty Values and Needs in Seven Disciplines* (Berkeley, CA: Center for Studies in Higher Education, 2010), https://escholarship.org/uc/cshe_fsc

but rather one form of dissemination among a broad range, or a form that becomes significant in its accretion, as a long-tended scholarly blog or Twitter account can do. At certain stages of the research process, it is often not as important to produce an in-depth scholarly summation so much as to provide short snapshots of an experiment's current developments (as in the hard sciences), or an analysis of a source (in the humanities). This is a situation where it may be more appropriate for a scholar to write small reports in the form of blog entries and publicise them on various social networks. Blogs offer a suitable platform for initial scholarly sharing, with both online availability and the possibility for commenting on the actual scholarly content; or, indeed, they can occupy one layer in a wider transmedial scholarly production. It is also a simple way to gain an audience for a specific result, or present observations step-by-step, for instance, during an archaeological campaign. Ideally such blogging occurs within a secure scholarly environment, such as Hypotheses.org,[39] where researchers benefit from editorial support as well as wide visibility. This epitomises the spirit of what one scholar has referred to as 'Open Notebook History'[40] and another as 'forking' history.[41] New hybrids able to harness such approaches also continue to appear, such as the PARTHENOS Hub[42] and the OpenMethods Metablog,[43] including those that offer wholly new forms of argumentation, such as the logicist publication format proposed for archaeology.[44] But even if a scholar were able to create and disseminate a trail of micro publications, many of which might be cited by peers as being interesting and useful knowledge, how could this coinage then be exchanged for those most valuable of assets: reputation, recognition, and professional advancement? How, indeed, would the author(s) be able to avoid the fate of the excellent *French Book Trade in*

---

39   *Hypotheses: Academic Blogs*, http://hypotheses.org/
40   W. Caleb McDaniel, 'Open Notebook History' (22 May 2013), http://wcm1.web.rice.edu/open-notebook-history.html
41   Konrad M. Lawson, 'Fork the Academy', ProfHacker, *The Chronicle of Higher Education* (30 April 2013), http://www.chronicle.com/blogs/profhacker/fork-the-academy/48935
42   'PARTHENOS Hub', *PARTHENOS*, http://www.parthenos-project.eu/portal/the-hub
43   *OpenMethods*, https://openmethods.dariah.eu
44   Pierre-Yves Buard et al. 'The Archaeological Excavation Report of Rigny: An Example of an Interoperable Logicist Publication', *CIDOC* (Heraklion, Greece: 2018), ffhal-01892412f.

Enlightenment Europe, released more as a resource than an argument, and subsequently subjected to critique based on its incompleteness and its potential to mislead users (something which analogue primary sources have the capacity to do as well).[45] It is in such cases that the effect of the second main issue that hinders the proliferation of new forms of publication can be clearly identified, that is, authority. Peer review has been, and will remain, the gold standard for proving academic quality for the foreseeable future. In fact:

> [c]onventional peer review is so central to scholars' perception of quality that its retention is essentially a sine qua non for any method of archival publication, new or old, to be effective and valued. Peer review is the hallmark of quality that results from external and independent valuation. It also functions as an effective means of winnowing the papers that a researcher needs to examine in the course of his or her research.[46]

Peer review remains both the essential foundation and a major barrier within the current scholarly communication system. The system is widely viewed as deeply flawed because of the time and expense it requires and its inherent potential for uneven results. In spite of this, it is still viewed as being greatly superior to any alternative; with such approaches as altmetrics[47] and bibliometric-driven impact factors coming under particular and sustained critique.[48] This does not mean that peer review cannot change, and cannot itself become more efficient and better suited to supporting the various sizes, shapes, and media forms of publication. New models such as open peer review 'manuscript marketplaces',[49]

---

45 The original resource can be found here: http://fbtee.uws.edu.au/main/, the critical review (by Robert Darnton) here: http://www.history.ac.uk/reviews/review/1355, and the 'critique of the critique' (by Mark Curran) here: https://doi.org/10.1017/s0018246x12000556

46 Diane Harley et al., 'The Influence of Academic Values on Scholarly Publication and Communication Practices', *Journal of Electronic Publishing*, 10.2 (2007), http://hdl.handle.net/2027/spo.3336451.0010.204

47 James Wilsdon et al., *Next-generation Metrics: Responsible Metrics and Evaluation for Open Science*, Report of the European Commission Expert Group on Altmetrics (Luxembourg: Publications Office of the European Union, 2017), https://ec.europa.eu/research/openscience/pdf/report.pdf

48 Diane Harley et al., *Assessing the Future Landscape*; Wilsdon, James, et al., *The Metric Tide: Report of the Independent Review of the Role of Metrics in Research Assessment and Management* (HEFCE: London, 2015), https://doi.org/10.13140/RG.2.1.4929.1363

49 G. Eysenbach, 'Peer-Review 2.0: Welcome to JMIR Preprints, an Open Peer-Review Marketplace for Scholarly Manuscripts', *JMIR Preprints*, 1.1 (2015), e1, https://doi.org/10.2196/preprints.5337

and other forms of review associated with overlay and review journals (including some specifically aimed at digital publication such as RIDE)[50] are being piloted. However, such new models will require not only technical platforms, but also a new 'social contract' between publishers, institutions, and researchers so as to provide a more democratic, but equally well run, system. The central position of publishers like Elsevier, based in large part on their management of the quality control system across a range of disciplines, is crumbling in the face of their increasing profits, and the increasing budgetary pressures on libraries as countries like Germany and Sweden take strong negotiation stances.[51] The fact that a journal like *Glossa*,[52] a journal of the publisher Open Library of the Humanities,[53] could be founded on the basis of the protest resignation of the entire editorial board of an Elsevier journal, also evidences the level of frustration on the production side of publication culture. Consumption-side negotiations grab fewer headlines as they tend to be individual rather than institutional, but certainly the gaining in popularity of open science (to be discussed in greater detail below), with its focus on publications as well as data, rewards, training, and ethics, indicates the form this new contract may take. The culture change that stands before the scholarly community to enable the acceptance of new publication modes must also include a negotiation of the meaning and value of the metrics and review mechanisms, and enable a re-evaluation of the many proxies upon which we still rely: from publishers to citations to alternative metrics.

Some aspects of this new review model may themselves occur by proxy: citations may not carry the same weight in humanities disciplines that they do in the sciences, but certainly a protected idea that is referenced widely will have proven its impact if not its quality. The challenge is not to divest ourselves of all that is a part of the tradition or all that is emerging in other disciplines, but to understand what it means for the humanities and to apply it appropriately. Plenty of electronic platforms and publishers have demonstrated viable and reliable practices for managing quality assessments that are overt as

---

50 RIDE, *A Review Journal for Digital Editions and Resources* (IDE), http://ride.i-d-e.de/
51 Holly Else, 'Dutch Publishing Giant Cuts off Researchers in Germany and Sweden', *Nature*, 559 (2018), 454–55 https://www.nature.com/articles/d41586-018-05754-1
52 *Glossa: A Journal of General Linguistics*, https://www.glossa-journal.org/
53 *Open Library of the Humanities*, https://www.openlibhums.org/

well as covert, though relatively few of them (only ten percent by one reckoning)[54] specifically target humanities and social sciences. The use of a platform like CommentPress[55] harnesses this impetus. However, impetus — along with the cost in time and effort in seeking reviews in multiple journals — may, at some point, make independent review options, like Publons,[56] more attractive, more utilised, and ultimately a viable and accepted pathway to validation. In such a system, user registration information might indicate academic expertise, as might community self-regulation, although the threat of incivility on such platforms must also be managed. The binary simplicity of Facebook 'friending' and 'liking' may therefore not be fit for this purpose. If a young scholar is able to document the positive responses to his or her work over a period of months or years from known, senior scholars in their field, then this should be captured and considered, as many of the 'next generation' metrics platforms and approaches now do (e.g. altmetrics).[57] If nothing else, it could be controlled by interest and active understanding, rather than by a formal loop regulated by a publisher who may reject good work, not on the basis of its quality, but rather because of externalities related to the focus of the press or the nature of their publications, such as a work's length, language, or format.

By transferring the editorial and curatorial functions to the researcher-users, some unique and useful formats for scholarship can arise within the humanities and its peripheries. At the most basic level, there is a wide but uneven provision of independent national and institutional research repositories that provide the most basic infrastructure for making research accessible without necessarily promoting its visibility or authority. In certain cases, this model can work well: the arXiv preprint repository,[58] for example, is now a cornerstone of physics research. More elaborate cognates also exist,

---

54 T. Ross-Hellauer, 'What is Open Peer Review? A Systematic Review', *F1000 Science Policy Research Gateway*, 6.588 (2017), https://doi.org/10.12688/f1000research.11369.2

55 The Institute for the Future of the Book, 'Welcome to CommentPress', *Future of the Book*, http://futureofthebook.org/commentpress/

56 In particular, community-based models like *Publons* (https://publons.com/home/) show promise in this field, although, one must be wary of editing services companies offering fee-based, non-specific peer review of scientific manuscripts as well.

57 *Altmetrics: Who's talking about your research?*, https://www.altmetric.com/

58 *arXiv.org*, https://arxiv.org/

such as the CARMEN Virtual Laboratory (VL) a 'cloud-based platform which allows neuroscientists to store, share, develop, execute, reproduce and publicise their work [... including] an interactive publications repository. This new facility allows users to link data and software to publications.'[59] More at the margins of traditional forms, perhaps, are publication outlets like JoVE[60] (which publishes research results in the form of short video clips), and conceptual approaches like 'explorable explanations',[61] which resist not only the traditional formats of the monograph, but also the tradition of the authorial voice, and presents instead the data underlying the author's conclusions and lets the readers develop their own interpretations. Projects like THOR are establishing interdisciplinary and inter-sectoral collaborations to create a 'seamless integration between articles, data, and researchers across the research lifecycle'.[62] Traditional publishers are entering the space as well, with new entrants such as Open Book Publishers experimenting with hybrid publications using Wikimedia Commons; established players opening up new platforms, such as Palgrave's open format Pivot platform;[63] and new, collaborative, funder-driven platforms for publication, such as the Wellcome/F1000 cooperative venture Wellcome Open Research.[64]

These platforms bring us back to the reception and adoption of open science principles, and the mixed reception the concept of open access has had in the humanities. Given the long publication cycles, the lack of reuse of results by industry, and the mix of books and articles found on the traditional humanist's publication record, the average humanistic researcher has perhaps felt at a distance from the push to

---

59 Victoria Jane Hodge et al., 'A Digital Repository and Execution Platform for Interactive Scholarly Publications in Neuroscience', *Neuroinformatics*, 14.1 (2016), 23–40 (p. 23), https://doi.org/10.1007/s12021-015-9276-3
60 *JoVE | Peer Reviewed Scientific Video Journal: Accelerating Scientific Research & Education*, https://www.jove.com
61 Maarten Lambrechts, 'The Rise of Explorable Explanations', *Maarten Lambrechts* (4 March 2015), http://www.maartenlambrechts.com/2015/03/04/the-rise-of-explorable-explanations.html
62 *Project THOR*, https://project-thor.eu/
63 Hazel Newton, 'Breaking Boundaries in Academic Publishing: Launching a New Format for Scholarly Research', *Insights*, 26.1 (2013), 70–76, https://doi.org/10.1629/2048-7754.26.1.70
64 Robert Kiley, 'Why We're Launching a New Publishing Platform', *Wellcome* (7 July 2016), https://wellcome.ac.uk/news/why-were-launching-new-publishing-platform

ensure public access to research, viewing it as 'good citizenship'[65] rather than a professional imperative. In addition, the fact that a majority of humanities research is developed without external research funding makes the discussion of the 'gold' access route ring particularly hollow: an average article processing charge (APC) in the sciences could well absorb a humanist's only access to a research budget and institutional contributions to research travel, for several years. Even the wide availability of 'green' deposit options does not resonate as it perhaps should, with the greater concern being the fate of research released as a digital edition or other free-standing form of scholarship without the benefit of oversight by a publisher. Anecdotally, one also hears of editorial boards giving preference to pieces not already in preprint, and of tenure committees expecting a book from the highest impact publisher, regardless of their publication policies. As the attention of funding agencies and national research agencies begins to focus on ensuring open access, however, one can expect the awareness and emphasis of openness to increase in the humanities. One has to expect that some aspects of this shift will require the 'stick' of possible sanctions to be applied in the cases of non-compliant researchers, but also that it will take advantage of the 'carrots' — personal, professional and informational — that wider dissemination can bring. This will be of particular importance in the digital humanities, where the traditional measured pace in humanities scholarship meets the rapid changes of technology head on: 'without free and open access to these materials, the majority of the innovations of the Digital Humanities will remain [...] a tremendously fascinating instrumentarium but the internet's genuinely transformational promise will have been missed, largely as a result of our failure to understand the full implications of the digital medium itself.'[66]

Even within traditional length formats of scholarly communication, if we are still attached to traditional forms of journal editing, we can observe that its core services, namely, identification, certification,

---

65 Sheila Anderson, 'What are Research Infrastructures', *International Journal of Humanities and Arts Computing*, 7.1–2 (2013), 4–23 (p. 5), https://doi.org/10.3366/ijhac.2013.0078

66 Sigi Jöttkandt, 'Free/Libre Scholarship: Open Humanities Press', unpublished Conference Paper at HumaniTech, UC Irvine, 3 April 2008, p. 6, http://eprints.rclis.org/3824/1/Jottkandt-03-april-08-Irvine-talk.pdf

dissemination, and long-term availability, can be implemented on the basis of an existing publication repository. Indeed, such a repository can provide a submission environment that identifies authors and time-stamps the document, and offers a perfect online dissemination platform with the necessary long-term archiving facility of the hosting institution. In such a context, designing a certification environment mechanism whereby a paper deposited by an author is forwarded to an editorial committee for peer review is quite a straightforward endeavour. This is exactly what is now being experimented with by the Episciences.org[67] project on top of the French Hyper Articles en Ligne (HAL) open repository platform.[68] This platform is further interesting in that it offers new possibilities for changing our perspective on the certification process: open submission, open peer review,[69] updated versions of articles, and community feedback are features that may dramatically change our views of scholarly publishing.

Research Data and the Evolving Communications Landscape

The unstated implication of many of these innovations is not only that the publication should appear in a range of places and a range of formats to meet both the needs of authors and readers, but also that the publication should make research data and the research process explicit, not only the research results. Seen from this perspective, a platform such as CommentPress,[70] which exposes the formation of peer opinion around a work of scholarship in real time, should also inhabit a place along this continuum. If this perspective were to be advanced toward its natural conclusion, a number of interesting avenues for sharing scholarship could be opened up. Developing objects of scholarship that are able to expose a full epistemological process, rather than a summation of its conclusions, would enable scholars to access the output of others in a more holistic, organic fashion, and reduce some of the requirements for

---

67 *Episciences.org*, https://www.episciences.org/
68 *HAL Open Repository Platform*, https://hal.archives-ouvertes.fr
69 Tom DeCoursey, 'The Pros and Cons of Open Peer Review', *Nature* (2006), https://doi.org/10.1038/nature04991
70 The Institute for the Future of the Book, 'Welcome to CommentPress', *Future of the Book*, http://futureofthebook.org/commentpress/

authors to use their text to justify a place in the state of the art or make general conclusions about a field of study (something not every reader may want or need). In addition, such a publication — which could in theory have a much more open format and a more variable length than a book or even a traditional journal article — could accelerate the capacity for the humanistic knowledge ecosystem to share and exchange information, thus reducing the likelihood of competing work being developed in parallel and increasing the potential for the identification of shared interests and fostering of collaboration. Again, models exist in the sciences and at the edges of the humanities, where in archaeology or, indeed, the biosciences, a discovery may be recorded and made public with only a short observation or note to contextualise it. This should be possible in disciplines such as literature and history as well, not to mention formats such as those developed by the Centre Virtuel de la Connaissance sur l'Europe,[71] or the CENDARI project's Archival Research Guide's[72] gesture toward how this sort of exchange might occur. This kind of 'light-touch' format is valuable not only for its flexibility and potential technical integration (for example, via services to uplift and expose significant named entities within and across works, thereby enhancing visibility in a targeted way for research), but also for the visibility it can bring to less-established scholars, or para-academics in 'alt-ac' (alternative-academic) style roles, or to work that is not best presented in one of the traditional formats.

Conclusions

The CIBER study cited above also asked the question: what will the information environment be like in 2017? Having now passed that landmark, it is uncanny how much of the report's speculation still seems germane: for example, the suspicion that research processes and publications would need to change drastically to take advantage of the opportunities and respond to the current inequities in the scholarly publishing environment. However, it is not the technology at hand

---

71 'ePublications', *CVCE*, http://www.cvce.eu/en/epublications
72 'Intro to Thematic Research Guides', *Cendari*, http://www.cendari.eu/thematic-research-guides/intro-thematic-research-guides. CENDARI is a hybrid publication of undetermined length bringing together analysis, links to data sources, and semantic linking to related resources.

that needs to change for such a system of alternatives to conventional publishing to emerge, become normalised, and be accepted as works of scholarship. The book and the monograph will not disappear, nor should they; but the primacy of the book as the privileged format of humanities scholarship will need to cede some room to outputs that are more process- and sharing-oriented, or less prone to claims of representing the authority of the 'final word' merely because of their length or adherence to the expected proxies of look, feel, or publisher's branding. Scholarship will also benefit from recognising deep and sustained engagement with ideas across many publications and publication outlets: as Christine Borgman asserts, we must create an information infrastructure that supports scholarship in all its multiple forms of communication.[73] The potential basic unit of scholarship must be expanded to include not just the book, chapter, and article; but the scholar, the project, the team, and the career. Indeed, the growth in acceptance of the ORCID system[74] for identifying scholars may indicate a shift in this direction. The idea is also not to perpetuate a system in which word counts are arbitrarily constrained in order to achieve the smallest publishable unit (*Science* and *Nature* being the extreme examples), a practice that scholars have rightly criticised.[75] Instead, the 'science telescope', as it were, needs to be fitted with an adjustable magnification, which scholars may use as befits their findings and research process, if we are to accommodate the needs of those whose work may be interdisciplinary, transnational, and experimental. Harley et al's extensive study on scholarly communication bears this out as being one of their five primary recommendations and findings: it calls for '[n]ew models of publication that can accommodate arguments of varied length, rich media, and embedded links to data; plus institutional assistance to manage permissions of copyrighted material'. This would address the problem identified by that team: 'One of the biggest problems […] is that there is no clear understanding about what a digital or electronic equivalent of a book could be.'[76]

---

[73] Christine Borgman, *Scholarship in the Digital Age: Information, Infrastructure and the Internet* (Cambridge, MA: MIT, 2007).
[74] *ORCID: Connecting Research and Researchers*, https://orcid.org/
[75] Diane Harley et al., *Assessing the Future Landscape*, p. 442.
[76] Ibid.

This question of the functional aspects of the new scholarly communication is further supported by a second major requirement related to the forms these new publication models might take, which call for '[s]upport for managing and preserving new research methods and products, including components of natural-language processing, visualization, complex distributed databases, GIS, among many others'.[77] It is the culture of the institutions and the disciplines that need to stretch to accommodate these possibilities, to allow them to find a 'voice' that can support their transmission and validate their results. The developments in research infrastructure, like the platforms mentioned above, are ready to create such safe places for scholarship to extend its reach. However, for their impact to be felt, they must be met at institutional levels with enthusiasm and understanding rather than suspicion.

The various possibilities outlined so far only make sense if research institutions invest time, political capital, and budget to implement such models, and make them part of the daily life of their researchers. A typical example of best practice can be taken from the recently published open access policy by the INRIA[78] research institute, which combines the elements of a mandate to deposit all publications on the HAL archive, a cautious assessment of any new models provided by the private publishing sector, and the funding of the Episciences.org platform.

Adopting a less conservative vision of scholarly communication opens up a whole range of possibilities for improving the way scientific ideas can be seamlessly transmitted to a wide audience. We can see that a new landscape can be outlined where the management of virtual research environments, comprising research data, various types of notes and commentaries, as well as draft documents that link these objects together; could dramatically change the way scholarship is carried out in the future. The Dutch national data service Data Archiving and Networked Services (DANS) (in cooperation with Brill) is already harnessing this potential with their online journal for research data,[79] while DARIAH's ERIC is promoting a culture of greater sharing among researchers as well as between researchers and cultural heritage

---

77 Ibid, p. 20.
78 'Inria Champions Open Access', *Inria* (6 November 2015), https://www.inria.fr/en/news/news-from-inria/inria-champions-open-access
79 'DANS and Brill Publishers Launch Online Journal on Research Data', *DANS* (20 October 2015), https://dans.knaw.nl/en/current/news/dans-and-brill-publishers-launch-online-journal-on-research-data

institutions.[80] In such environments, various levels of peer review are possible, from simple feedback by known colleagues, to the possibility for any member of a research community to comment at length. Traditional peer review is just one possible implementation of such a model where the main objective should be, as it has always been, to improve quality and widen accessibility for new ideas and the output of research, and to rebalance the values we communicate through the way we use scholarship with those expressed by our dissemination and communication infrastructures.

Bibliography

'About', *Open Methods*, https://openmethods.dariah.eu/about/

Adema, Janneke and Paul Rutten, 'Digital Monographs in the Humanities and Social Sciences: Report on User Needs', *OAPEN Deliverable*, 3.1.5 (2010), https://openreflections.files.wordpress.com/2008/10/d315-user-needs-report.pdf

Altmetrics: Who's talking about your research?, https://www.altmetric.com/

Anderson, Sheila, 'What are Research Infrastructures', *International Journal of Humanities and Arts Computing*, 7.1–2 (2013), 4–23, https://doi.org/10.3366/ijhac.2013.0078

arXiv.org, https://arxiv.org/

'Back to Books', *Nature*, 463 (2010), 588–88, https://doi.org/10.1038/463588a

Borgman, Christine, *Scholarship in the Digital Age: Information, Infrastructure and the Internet* (Cambridge MA: MIT Press, 2007).

Buard, Pierre-Yves, et al., 'The Archaeological Excavation Report of Rigny: An Example of an Interoperable Logicist Publication', unpublished conference paper at *CIDOC*, Heraklion, Greece: 2018, https://hal.archives-ouvertes.fr/hal-01892412/document

Bulger, Monica, et al., *Reinventing Research? Information Practices in the Humanities*, Research Information Network Report (London: The Research Information Network, 2011).

Catalano, Frank, 'Paper is Back: Why "Real" Books Are on the Rebound', *GeekWire* (18 January 2015), http://www.geekwire.com/2015/paper-back-real-books-rebound/

---

80   Laurent Romary et al., 'Data Fluidity in DARIAH: Pushing the Agenda Forward', *BIBLIOTHEK Forschung und Praxis*, 39.3 (2016), 350–57, https://doi.org/10.1515/bfp-2016-0039, https://hal.inria.fr/hal-01285917v2

Cheverie, Joan, Jennifer Boettcher, and John Buschman, 'Digital Scholarship in the University Tenure and Promotion Process: A Report on the Sixth Scholarly Communication Symposium at Georgetown University Library', *Journal of Scholarly Publishing*, 40.3 (2009), 219–30, https://doi.org/10.3138/jsp.40.3.219

Crossick, Geoffrey, *Monographs and Open Access: A Report to HEFCE* (London: HEFCE, 2015), https://dera.ioe.ac.uk/21921/1/2014_monographs.pdf

Curran, Mark, 'Beyond the Forbidden Best-Sellers of Pre-Revolutionary France', *The Historical Journal*, 56.1 (2013), 89–112, https://doi.org/10.1017/s0018246x12000556

'DANS and Brill Publishers Launch Online Journal on Research Data', *DANS* (20 October 2015), https://dans.knaw.nl/en/current/news/dans-and-brill-publishers-launch-online-journal-on-research-data

Darnton, Robert, 'Review of The French Book Trade in Enlightenment Europe, 1769–1794', *Reviews in History* (December 2012), http://www.history.ac.uk/reviews/review/1355

DeCoursey, Tom, 'The Pros and Cons of Open Peer Review', *Nature* (2006), https://doi.org/10.1038/nature04991

Denbo, Seth, 'Draft Guidelines on the Evaluation of Digital Scholarship', *American Historical Association* (21 April 2015), http://blog.historians.org/2015/04/draft-guidelines-evaluation-digital-scholarship/

Else, Holly, 'Dutch Publishing Giant Cuts off Researchers in Germany and Sweden', *Nature*, 559 (2018), 454–55, https://www.nature.com/articles/d41586-018-05754-1

Engels, Tim C. E., et al., 'Are Book Publications Disappearing from Scholarly Communication in the Social Sciences and Humanities?', *Aslib Journal of Information Management*, 70.6 (2018), 592–607, https://doi.org/10.1108/AJIM-05-2018-0127

Episciences.org, https://www.episciences.org/

'ePublications', *CVCE*, http://www.cvce.eu/en/epublications

Estelle, Lorraine, 'What Researchers Told Us about their Experiences and Expectations of Scholarly Communications Ecosystems', *Insights*, 30.1 (2017), 71–75, https://doi.org/10.1629/uksg.349

Eysenbach, G., 'Peer-Review 2.0: Welcome to JMIR Preprints, an Open Peer-Review Marketplace for Scholarly Manuscripts', *JMIR Preprints*, 1.1 (2015), 1–1, https://doi.org/10.2196/preprints.5337

FBTEE: The French Book Trade in Enlightenment Europe | Mapping the Trade of the Société Typographique de Neuchâtel, 1769-1794, http://fbtee.uws.edu.au/main/

Feinberg, Melanie, 'Beyond Digital and Physical Objects: The Intellectual Work as a Concept of Interest for HCI', in *Proceedings of the SIGCHI Conference on*

Human Factors in Computing Systems (Paris, France, April 27-May 02, 2013), pp. 3317–26, https://doi.org/0.1145/2470654.2466453

Giminez-Toledo, Elea, et al., 'The Evaluation of Scholarly Books as Research Output: Current Developments in Europe', unpublished conference paper at the *15th International Society for Scientometrics and Informetrics Conference*, Istanbul, Turkey, 29th June–4th July 2015, http://curis.ku.dk/ws/files/141056396/Giminez_Toledo_etal.pdf

Glossa: A Journal of General Linguistics, https://www.glossa-journal.org/

Greenblatt, Stephen, 'A Special Letter from Stephen Greenblatt', *Modern Language Association* (28 May 2002), https://www.mla.org/Resources/Research/Surveys-Reports-and-Other-Documents/Publishing-and-Scholarship/Call-for-Action-on-Problems-in-Scholarly-Book-Publishing/A-Special-Letter-from-Stephen-Greenblatt

Guillory, John, 'How Scholars Read', *ADE Bulletin*, 146 (2008), 8–17, https://doi.org/10.1632/ade.146.8

HAL Open Repository Platform, https://hal.archives-ouvertes.fr

Halliday, Leah, 'Scholarly Communication, Scholarly Publication and the Status of Emerging Formats', *Information Research*, 6.4 (2001), http://www.informationr.net/ir/6-4/paper111.html

Harley, Diane, et al., *Assessing the Future Landscape of Scholarly Communication: An Exploration of Faculty Values and Needs in Seven Disciplines* (Berkeley, CA: Center for Studies in Higher Education, 2010), https://escholarship.org/uc/cshe_fsc

Harley, Diane, et al., 'The Influence of Academic Values on Scholarly Publication and Communication Practices', *Journal of Electronic Publishing*, 10.2 (2007), https://doi.org/10.3998/3336451.0010.204, http://hdl.handle.net/2027/spo.3336451.0010.204

Hodge, Victoria Jane, et al., 'A Digital Repository and Execution Platform for Interactive Scholarly Publications in Neuroscience', *Neuroinformatics*, 14.1 (2016), 23–40, https://doi.org/10.1007/s12021-015-9276-3

Holzman, Alex, and Robert Brown, eds., 'Special Issue on Digital Publishing for the Humanities and Social Sciences', *Journal of Scholarly Publishing*, 48.2 (2017), https://doi.org/10.3138/jsp.48.2.73

Hypotheses: Academic Blogs, http://hypotheses.org/

'Inria Champions Open Access', *Inria* (6 November 2015), https://www.inria.fr/en/news/news-from-inria/inria-champions-open-access

'Intro to Thematic Research Guides', *Cendari*, http://www.cendari.eu/thematic-research-guides/intro-thematic-research-guides

Jones, Jacqueline, 'AHA Statement on Policies Regarding the Embargoing of Completed History PhD Dissertations', *American Historical Association*

(22 July 2013), http://blog.historians.org/2013/07/american-historical-association-statement-on-policies-regarding-the-embargoing-of-completed-history-phd-dissertations/

Jöttkandt, Sigi, 'Free/Libre Scholarship: Open Humanities Press', unpublished Conference Paper at HumaniTech, UC Irvine, 3 April 2008, http://eprints.rclis.org/3824/1/Jottkandt-03-april-08-Irvine-talk.pdf

JoVE | Peer Reviewed Scientific Video Journal: Accelerating Scientific Research & Education, https://www.jove.com

Kiefer, Ferenc, 'ERIH's Role in the Evaluation of Research Achievements in the Humanities', in *New Publication Cultures in the Humanities Exploring the Paradigm Shift*, ed. by Péter Dávidházi (Amsterdam: Amsterdam University Press, 2014), pp. 173–82, http://www.oapen.org/search?identifier=515678

Kiley, Robert, 'Why We're Launching a New Publishing Platform', *Wellcome* (7 July 2016), https://wellcome.ac.uk/news/why-were-launching-new-publishing-platform

Lambrechts, Maarten, 'The Rise of Explorable Explanations', *Maarten Lambrechts* (4 March 2015), http://www.maartenlambrechts.com/2015/03/04/the-rise-of-explorable-explanations.html

Lawson, Konrad M., 'Fork the Academy', ProfHacker, *The Chronicle of Higher Education* (30 April 2013), http://www.chronicle.com/blogs/profhacker/fork-the-academy/48935

Leech, Dennis, 'Perverse Incentives Mean the REF Encourages Mediocrity rather than Excellence', *REF Watch* (10 December 2013), http://ref.web.ucu.org.uk/2013/12/10/perverse-incentives-mean-the-ref-encourages-mediocrity-rather-than-excellence/

Lyons, Rebecca, and Samantha Rayner, eds., *The Academic Book of the Future* (Basingstoke: Palgrave Macmillan, 2016), https://doi.org/10.1057/9781137595775

McCarty, Willard, *Humanities Computing* (Basingstoke: Palgrave Macmillan, 2005).

McDaniel, W. Caleb, 'Open Notebook History' (22 May 2013), http://wcm1.web.rice.edu/open-notebook-history.html

Millen, Caren, 'Exploring Open Access to Save Monographs, the Question Is — How?', *Jisc*, https://www.jisc.ac.uk/blog/exploring-open-access-to-save-monographs-the-question-is-how-24-oct-2012

Miller, Daniel, *The Comfort of Things* (Malden, MA: Polity, 2008).

Milliot, Jim, 'For Books, Print Is Back', *PublishersWeekly.com* (2 January 2015) https://www.publishersweekly.com/pw/by-topic/industry-news/bookselling/article/65172-print-is-back.html

Naughton, John, 'The Future of News (and of Lots More Besides), *Memex 1.1* (17 March 2009), http://memex.naughtons.org/archives/2009/03/17/6998

Newton, Hazel, 'Breaking Boundaries in Academic Publishing: Launching a New Format for Scholarly Research', *Insights*, 26.1 (2013), 70–76, https://doi.org/10.1629/2048-7754.26.1.70

Open Library of the Humanities, https://www.openlibhums.org/

ORCID, Connecting Research and Researchers, https://orcid.org/

'PARTHENOS Hub', *PARTHENOS*, http://www.parthenos-project.eu/portal/the-hub

Project THOR, https://project-thor.eu/

Publons, https://publons.com/home/

RIDE: A Review Journal for Digital Editions and Resources (IDE), http://ride.i-d-e.de/

Romary, Laurent, et al., 'Data Fluidity in DARIAH — Pushing the Agenda Forward', *BIBLIOTHEK Forschung und Praxis*, 39.3 (2016), 350–57, https://doi.org/10.1515/bfp-2016-0039, https://hal.inria.fr/hal-01285917v2

Ross-Hellauer, T., 'What is Open Peer Review? A Systematic Review', *F1000 Science Policy Research Gateway*, 6.588 (2017), https://doi.org/10.12688/f1000research.11369.2

Ian Rowlands et al., 'Information Behaviour of the Researcher of the Future', *CIBER Briefing Papers* (2008), https://www2.warwick.ac.uk/study/cll/courses/professionaldevelopment/wmcett/researchprojects/dialogue/the_google_generation.pdf

Shirky, Clay, 'Newspapers and Thinking the Unthinkable', *Clay Shirky* (13 March 2009), http://www.shirky.com/weblog/2009/03/newspapers-and-thinking-the-unthinkable/

Shoemaker, Bob, 'The Future of the (e)Book', *History Matters* (1 December 2015), http://www.historymatters.group.shef.ac.uk/future-ebook/

Steele, Colin, 'Scholarly Monograph Publishing in the 21st Century: The Future More Than Ever Should Be an Open Book', *Journal of Electronic Publishing*, 11.2 (2008), https://doi.org/10.3998/3336451.0011.201

'Survey of Use of Monographs by Academics — as Authors and Readers', *OAPEN-UK* (2014), http://oapen-uk.jiscebooks.org/files/2012/02/OAPEN-UK-researcher-survey-final.pdf

The Institute for the Future of the Book, 'Welcome to CommentPress', *Future of the Book*, http://futureofthebook.org/commentpress/

'Total Number of Websites: Internet Live Stats', *Internet Live Stats*, http://www.internetlivestats.com/total-number-of-websites/

Truschke, A., 'Dissertation Embargoes and Publishing Fears', *Dissertation Reviews* (1 April 2015), http://dissertationreviews.org/archives/11842

Valauskas, Edward J., 'Waiting for Thomas Kuhn: First Monday and the Evolution of Electronic Journals', *First Monday*, 2.12 (1997), http://firstmonday.org/ojs/index.php/fm/article/view/567

Waters, Lindsay, 'A Modest Proposal for Preventing the Books of the Members of the MLA from Being a Burden to their Authors, Publishers, or Audiences', *PMLA*, 115.3 (2000), 315–17, https://doi.org/10.2307/463452.

Wilsdon, James, et al., *Next-generation Metrics: Responsible Metrics and Evaluation for Open Science*, Report of the European Commission Expert Group on Altmetrics (Luxembourg: Publications Office of the European Union, 2017), https://ec.europa.eu/research/openscience/pdf/report.pdf

Wilsdon, James, et al., *The Metric Tide: Report of the Independent Review of the Role of Metrics in Research Assessment and Management* (HEFCE: London, 2015), https://doi.org/10.13140/RG.2.1.4929.1363

Wolfe Thompson, Jennifer, 'The Death of the Scholarly Monograph in the Humanities? Citation Patterns in Literary Scholarship', *Libri*, 52.3 (2002), 121–36, https://doi.org/10.1515/LIBR.2002.121

4. The Impact of Digital Resources

Claire Warwick and Claire Bailey-Ross

It has now become commonplace to begin articles about the use and impact of digital resources with a bold statement about how much is being spent on their production. And it is a great deal, and seems to rise every year. What is less clear is exactly why we are doing this. We are often told that if it is not digital or digitised, it does not exist, and that this is especially true for our students. The corollary of this is the assumption that if things are digital, they not only exist, but are popular, exciting, well known, and thus well used. University managers, funding councils, and policy makers also appear to assume that doing things with computers is automatically better, faster, cheaper, and more economical in terms of person-time than not doing it, despite the lack of evidence for this.

It is no wonder, then, that there often seems to be an implied belief that doing humanities in a digital way will render it 'relevant', solve any apparent crises in the subject, and bring what has otherwise been obscure and arcane to the notice, and indeed love, of the general public. At the same time, cultural heritage organisations, such as museums, galleries, archives, and libraries, have been investigating ways in which they can use digital methods and social media as a vector for outreach and a way to increase visitor engagement.

But are these assumptions well founded? Do we render the humanities relevant simply by being digital? Do visitors automatically find it easier to engage with cultural heritage, with galleries, libraries, museums, or archives (GLAM) if the material is digitised? These questions fall into the realm of what has become known as impact assessment, whether it is carried out by government bodies, or by cultural heritage institutions themselves.[1] In the following chapter we examine the question of how digital resources might have an impact, and upon whom, in what way, and how it might be measured. We will also examine the necessary conditions for a resource or collection to have an impact, foremost among which is its continued existence — an obvious and necessary condition, but not necessarily one as easily achieved as might be expected.[2]

Understanding and Measuring Impact

The process of understanding and measuring impact (impact assessment) has many definitions, depending on the context in which it is used. There are well-established fields of impact assessment, such as environment, health, economic, and social impact assessment; but these have not normally been associated with humanities research or cultural heritage institutions, particularly with regard to digital content, collections, and resources.[3] Recent research into the value and impact of digitised resources and collections has shown clear benefits; but while there is an abundance of anecdotal evidence, systematic data is often lacking.[4]

For much of the last two decades the GLAM sector has taken the lead in measuring the impact of both its digital and physical collections. There has been a growing recognition that demonstrating, monitoring, and clearly articulating the impact and value of their existence is necessary in a time of intense pressure on public funding. Since the 1980s, the value and use of GLAM sector collections has been demonstrated through the lens of their 'impact', whether economic or social.[5]

---

1   Simon Tanner, *Measuring the Impact of Digital Resources: The Balanced Value Impact Model* (London: King's College London, 2012).
2   See, for example, James Smithies et al.,'Managing 100 Digital Humanities Projects: Digital Scholarship & Archiving in King's Digital Lab', *Digital Humanities Quarterly*, 13.1 (2019), http://www.digitalhumanities.org/dhq/vol/13/1/000411/000411.html
3   Sara Selwood, 'What Difference Do Museums Make? Producing Evidence on the Impact of Museums', *Critical Quarterly*, 44.4 (2002), 65–81, https://doi.org/10.1111/1467-8705.00457; Caroline Wavell et al., *Impact Evaluation of Museums, Archives and Libraries: Available Evidence Project* (Aberdeen: Robert Gordon University, 2002).
4   Simon Tanner, and Marilyn Deegan, *Inspiring Research, Inspiring Scholarship. The Value and Benefits of Digitised Resources for Learning, Teaching, Research and Enjoyment* (London: JISC, 2011).
5   John Myerscough, *The Economic Importance of the Arts in Britain* (London: Policy Studies Institute, 1988); Tony Travers, *Museums and Galleries in Britain Economic,*

Over the last fifteen years, a large amount of work has gone into forming and testing appropriate, flexible, and effective methodologies to indicate the impact and value of the GLAM sector. These include measuring attendance and demographics, audience evaluation, generic learning outcomes, and most recently, culture metrics.[6] For example, comprehensive monthly quantitative data is collected by all Department for Culture, Media, and Sport (DCMS)-sponsored museums and galleries in an attempt to reflect the quality and effectiveness of the programmes and the impact they have on society.[7] They provide a broad picture of performance with a focus on visitor figures, audience profiles, learning, outreach, visitor satisfaction, and income generation.

Although the frequency of evaluation is rising, whether it is meaningful in terms of its significance to long-term institutional impact assessment is still questionable, particularly in relation to digital resources. There is a need to address the 'use', 'value', and 'impact' of digital resources in the context of an expanding mass of cultural heritage digital content, which is believed to have tremendous potential for public engagement.

Current evaluation models, which are mainly project-driven, lack the consistency and longevity to create meaningful performance indicators and benchmarks. Many of the impact studies of museum and cultural

---

Social and Creative Impacts (London: London School of Economics & Political Science, 2006); François Matarasso, *Use or Ornament? The Social Impact of Participation in the Arts* (Stroud: Comedia, 1997); Naomi Kinghorn and Ken Willis, 'Measuring Museum Visitor Preferences Towards Opportunities for Developing Social Capital: An Application of a Choice Experiment to the Discovery Museum', *International Journal of Heritage Studies*, 14.6 (2008), 555–72, https://doi.org/10.1080/13527250802503290

6   Eilean Hooper-Greenhill, 'Measuring Learning Outcomes in Museums, Archives and Libraries: The Learning Impact Research Project (LIRP)', *International Journal of Heritage Studies*, 10.2 (2004), 151–74, https://doi.org/10.1080/1352725041000 1692877; *Culture Metrics: A Shared Approach to Measuring Quality*, http://www.culturemetricsresearch.com/

7   Department for Culture, Media, and Sport, *Statistical Data Set: Museums and Galleries Monthly Visits* (London, 2017). The Department for Culture, Media, and Sport (DCMS) sponsors sixteen national museums, which provide free entry to their permanent collections. These museums are the British Museum, Geffrye Museum, Horniman Museum, Imperial War Museum, National Gallery, National Maritime Museum, National Museums Liverpool, Science Museum Group, National Portrait Gallery, Natural History Museum, Royal Armouries, Sir John Soane's Museum, Tate Galleries, Tyne and Wear Museums, Victoria and Albert Museum, and the Wallace Collection. Data collection methods vary between institutions, and each uses a method appropriate to its situation. All data is collected according to the DCMS performance indicator guidelines.

activities overstate their measurable economic values but ignore the intangible impacts and values that they generate. Hasan Bakhshi and David Throsby, writing in 2010, believe that '[f]resh thinking is needed on how to articulate and, where possible, measure, the full range of benefits that arise from the work of arts and cultural organisations'.[8] However, this will be difficult; cultural impacts are often intangible, are more complex than the purely economic and numerical, and hard to explain and prove.[9] Visitor experience and engagement cannot be measured by instrumental values alone. As more collections are made available via digital technologies, the number of beneficiaries will increase and the ability of the sector to track and trace the benefits and end uses of visitor engagement with collections will become increasingly challenging.

The rise of 'impact' as an important concept in academic research, and the use of digital resources created in academia, is more recent. The LAIRAH (Log Analysis of Internet Resources in the Arts and Humanities) study found that very few creators of digital resources knew how they were used and had no contact with their user base.[10] Even funding bodies lacked knowledge about this; as Simon Tanner points out, LAIRAH was one of the first studies commissioned by the Arts and Humanities Research Council (AHRC) into the use of its resources.[11] However, in the twelve years since this study, changes are being made. Jisc became aware that investment in digital resources might be more strategically targeted, and so mandated user consultation and involvement in its second phase digitisation projects and commissioned a study, which resulted in the TIDSR (Toolkit for the Impact of Digital Scholarly Resources).[12] It proposed a number of different methods for evaluating the use of a digital resource.[13] This was a welcome development, but,

---

8   Hasan Bakhshi and David Throsby, *Culture of Innovation. An Economic Analysis of Innovation in Arts and Cultural Organizations* (Nesta, London, 2010), p. 58.
9   Wavell et al., *Impact Evaluation*.
10 Claire Warwick et al., 'If You Build It Will They Come? The LAIRAH Study: Quantifying the Use of Online Resources in the Arts and Humanities through Statistical Analysis of User Log Data', *Literary and Linguist Computing*, 23.1 (2008), 85–102, https://doi.org/10.1093/llc/fqm045
11 Tanner, *Measuring the Impact of Digital Resources*.
12 'TIDSR: Toolkit for the Impact of Digitised Scholarly Resources', *Oxford Internet Institute*, https://www.oii.ox.ac.uk/research/projects/tidsr/
13 Paola Marchionni, 'Why Are Users So Useful? User Engagement and the Experience of the JISC Digitisation Programme', *Ariadne* (30 October 2009), http://www.ariadne.ac.uk/issue/61/marchionni/

at the time, the idea of digital impact was associated only with use, findability, and dissemination: the toolkit involves such methods as web metrics, log analysis, surveys, focus groups, and interviews.

There is a strong underlying assumption, therefore, that use equals impact. The TIDSR team stresses that this is the reason for including qualitative techniques such as focus groups, because metrics may tell us how many people have landed on a certain page, or how many links are made to it; but they cannot tell us what the user thinks about what they have found, what they like and dislike, what they wanted or did not want, or, crucially, if they found what they were looking for. The toolkit was designed not only to provide evidence of use for the funders and institutions themselves, but also to help designers improve the resources; its utility has been proven in published studies such as those by Lorna M. Hughes et al.[14]

However, a major change in the idea of impact measurement occurred after TIDSR was produced: the UK's Research Excellence Framework (REF) adopted the idea of impact. The primary purpose of REF 2014 was to assess the quality of research in the UK's Higher Education Institutions (HEIs). A significant difference between the RAE (Research Assessment Exercise), last carried out in 2008, and REF 2014 was the inclusion of the assessment of impact.[15] This was defined as 'any effect on, change or benefit to the economy, society, culture, public policy or services, health, the environment, or quality of life, beyond academia'.[16] Under the terms of the REF, the conflation of use and simple dissemination of results was no longer acceptable. Academics now had to prove that their work had produced a change in behaviour of, or benefit to, a user community, and assessors were mandated to

---

14 Lorna M. Hughes et al., 'Assessing and Measuring Impact of a Digital Collection in the Humanities: An Analysis of the SPHERE (Stormont Parliamentary Hansards: Embedded in Research and Education) Project', *Digital Scholarship in the Humanities*, 30.2 (2015), 183–98, https://doi.org/10.1093/llc/fqt054

15 Molly Morgan Jones and Jonathan Grant, 'Making the Grade: Methodologies for Assessing and Evidencing Research Impact', in *7 Essays on Impact. DESCRIBE Project Report for Jisc*, ed. by David Cope et al. (Exeter: University of Exeter, 2013), pp. 25–43; Higher Education Funding Council of England (HEFCE), *The Nature, Scale, and Beneficiaries of Research Impact: An Initial Analysis of Research Excellence Framework (REF) 2014 Impact Case Studies* (London: King's College London, 2015).

16 REF, *Assessment Framework and Guidance on Submissions* (Bristol: REF UK, 2011), http://www.ref.ac.uk/2014/media/ref/content/pub/assessmentframeworkandguidanceonsubmissions/GOS%20including%20addendum.pdf

evaluate the reach and significance of such changes on a four-star scale. However, such effects are not straightforward to measure.

Impact evaluation is a complex issue, which is not helped by the fact that definitions are still being determined and understood by the sector. While there is an abundance of anecdotal evidence and descriptions of best practice, extensive evidence of impact, gathered systematically, is often lacking. The concept of impact is problematic because it is often entwined with several other key issues inherent in digital resources: discoverability, access, usage, and sustainability.[17] Considering the nature of these interwoven issues, is it possible to identify and measure impact in humanities research, particularly focusing on digital resources?

Sara Selwood suggests there are various ways of ascertaining, if not assessing, overall impact other than by economic value.[18] These include: direct consultation to assess public value; self-evaluations, and peer and user reviews; and stakeholder analysis.[19] Indeed, an increasing body of work is being developed around such approaches; but, to date, this has largely relied on peer and specialist review, which draws on small, professional networks rather than end-users.

Tanner has produced a complex model of impact assessment for GLAM institutions, which also defines impact as going beyond use to include benefit and change.[20] It takes into account multiple factors such as the ecosystem of a digital resource, the value drivers, and the key criteria indicators, all applied through five core functional stages: 1) context, 2) analysis and design, 3) implementation, 4) outcomes and results, and 5) review and respond; and it is evident that undertaking such an analysis would be a complex, time-consuming, and costly exercise.

---

17 Ben Showers, 'A Strategic Approach to the Understanding and Evaluation of Impact', in *Evaluating and Measuring the Value, Use and Impact of Digital Collections*, ed. by Lorna M. Hughes (London: Facet, 2012), pp. 63–72, https://doi.org/10.29085/9781856049085.006

18 Sara Selwood, 'Making a Difference: The Cultural Impact of Museums. An Essay for NMDC' (2010), https://www.nationalmuseums.org.uk/media/documents/publications/cultural_impact_final.pdf

19 Emily Keaney, 'Public Value and the Arts: Literature Review', *Strategy* (2006), 1–49 (p. 41); J. Holden and J. Baltà, *The Public Value of Culture: A Literature Review* (EENC Paper, Brussels, 2012).

20 Simon Tanner, 'The Value and Impact of Digitized Resources for Learning, Teaching, Research and Enjoyment', in *Evaluating and Measuring the Value, Use and Impact of Digital Collections*, ed. by Lorna M. Hughes (London: Facet, 2012), pp. 103–20, https://doi.org/10.29085/9781856049085.009

Nevertheless, we still lack adequate means to assess impact in humanities research due to a dearth of significant evidence beyond the anecdotal.[21] Despite the mass of existing evidence, 'attempts to interpret such evidence often tends (sic) to rely on assumptions about the nature of digital resources, without fully appreciating the actual way in which end users interact with digital content'.[22]

It is tempting draw a distinction, as Nancy Maron et al. do, between digital resources that are created by academics as part of their research, and the digitisation of collections and resources by GLAM institutions.[23] We might argue that the process of digitising a collection of papers, images, or museum objects for use by a memory institution differs from an academic, or group of academics, creating a digital resource as part of their research. It might be regarded as a service that is provided for the visiting public by the institution. It may be at least partially funded by the institution, and thus amenable to a more centralised, controlled process, and likely to be attached to an existing catalogue, or similar finding aid. An academic resource may be a piece of 'private enterprise' resulting from the individual's research interests. It is likely to be externally funded for a limited period, and may be somewhat idiosyncratic in design (this is more likely the older the resource is). In a large university, there may be numerous different homes for such projects: departments, computing centres, libraries, research units, digital humanities (DH) centres, or a combination of the above. In this way, the digital landscape may look, at least outwardly, more chaotic.

But this would be to oversimplify things. Many of the most celebrated digital research projects created by academics have resulted in very comprehensive digital resources, often known as archives (the Rossetti Archive,[24] the Blake Archive,[25] the Whitman Archive,[26] to name only a few), or in databases with huge, diverse user communities,

---

21 Ibid.
22 Tanner, *Measuring the Impact*, p. 23.
23 Nancy L. Maron, Jason Yun, and Sarah Pickle, 'Sustaining our Digital Future: Institutional Strategies for Digital Content', *Strategic Content Alliance*, Ithaka Case Studies in Sustainability (2013), https://sca.jiscinvolve.org/wp/files/2013/01/Sustaining-our-digital-future-FINAL-31.pdf
24 *Rossetti Archive*, www.rossettiarchive.org
25 *Blake Archive*, www.blakearchive.org
26 *Whitman Archive*, www.whitmanarchive.org

such as the Old Bailey Online.[27] Yet, they are the product of very complex and intellectually rigorous research, which could have, and in some cases has, resulted in the production of more traditional scholarly outputs such as articles and monographs.[28] It would also be a serious under-estimation to imply, in an age of highly skilled 'alt-ac' (alternative-academic) DH professionals working in museums, libraries, and archives, that resources created by GLAM institutions are simply about service and not the outcome of research. Tanner's model is designed for the GLAM sector, but draws explicitly on the definition of impact created for an academically driven exercise — the REF — and the process and model that he describes could easily be applied to an academically generated resource.

Digital resources may also have academic impact when a resource has an influence on the work of other academics. In the case of analogue resources, citations are commonly used as evidence of this; however, as Hughes et al., show, this is problematic in the case of digital resources, which are often not cited correctly.[29] Even in the case of conventional publications there are still significant problems in the use of metrics to judge academic impact and value: academics may cite papers as a straw man argument or an example of bad practice, and may cite in very different ways according to discipline — especially in the arts and humanities.[30] The gender of the author has also been proven to affect citation practices.[31] Thus, the most recent report concludes that metrics are not subtle enough to judge the quality of any kind of academic output, whether conventional or digital.[32]

---

27   *Old Bailey Online*, www.oldbaileyonline.org
28   Claire Warwick, 'Archive 360: The Walt Whitman Archive', *Archive Journal*, 1.1 (2011).
29   Hughes et al., 'Assessing and Measuring Impact'.
30   Björn Hellqvist, 'Referencing in the Humanities and its Implications for Citation Analysis', *Journal of the American Society for Information Science and Technology*, 61.2 (2010), 310–18, https://doi.org/10.1002/asi.21256
31   Daniel Maliniak, Ryan Powers, and Barbara F. Walter, 'The Gender Citation Gap in International Relations', *International Organization*, 67.4 (2013), 889–922, https://doi.org/10.1017/s0020818313000209; Jevin D. West et al., 'The Role of Gender in Scholarly Authorship', ed. by Lilach Hadany, PLOS ONE, 8.7 (2013), e66212, https://doi.org/10.1371/journal.pone.0066212
32   Wilsdon, James, et al., *The Metric Tide: Report of the Independent Review of the Role of Metrics in Research Assessment and Management* (HEFCE: London, 2015), https://doi.org/10.13140/RG.2.1.4929.1363

REF impact was assessed according to its reach and significance, and awarded star ratings from unclassified (little or no evidence of reach or significance) to four-star (outstanding).[33] Case studies also had to provide evidence for a link between this impact and the underpinning research, which had to be a two-star (internationally recognised) research output.[34] The case studies are now available in a database that, despite the caveats discussed above, provides useful evidence for the impact of UK research, whether digital or analogue. In the following section, we present a qualitative analysis of the impact of digital humanities as evidenced by the case study database. A previous quantitative text-mining-based study of all the REF case studies provides excellent evidence for the diversity of impacts claimed for research carried out in the UK's universities.[35] However, the report itself makes clear that this kind of method has limitations. Using text-mining methods, we can track the kinds of impact discussed: the words used, and the connections between themes and subject areas. This in itself is fascinating, but it provides only partial information. For example, case study authors claimed impact, but, the database does not indicate whether this claim was accepted by the panels as being wholly or partially evidenced, nor do we know how effective it was judged to be. Marks are released as a statistical profile across a unit, so we cannot link an individual case study to a star rating, unless all the case studies in that unit, from that university, were marked the same (which is relatively unusual). Nor do we know why the panel made the judgements they made, or how they marked reach and significance.

We therefore did not use text-mining methods, since this chapter is concerned primarily with exploring the types and quality of impact produced by DH, and the arguments that may be made for it. Instead,

---

33 REF, 'Assessment Criteria and Level Definitions', https://www.ref.ac.uk/2014/panels/assessmentcriteriaandleveldefinitions/

34 For further details on REF and impact see: Rita Marcella, Hayley Lockerbie, and Lyndsay Bloice, 'Beyond REF 2014: The Impact of Impact Assessment on the Future of Information Research', *Journal of Information Science*, 42.3 (2016), 369–85, https://doi.org/10.1177/0165551516636291; Rita Marcella et al., 'The Effects of the Research Excellence Framework Research Impact Agenda on Early- and Mid-Career Researchers in Library and Information Science', *Journal of Information Science*, 44.5 (2018), 608–18, https://doi.org/10.1177/0165551517724685; Clare Wilkinson, 'Evidencing Impact: A Case Study of UK Academic Perspectives on Evidencing Research Impact', *Studies in Higher Education*, 44.1 (2019), 72–85, https://doi.org/10.1080/03075079.2017.1339028

35 HEFCE, *Nature, Scale, and Beneficiaries*.

we sampled those case studies that were likely to be most relevant to DH methods by means of a phrase search for 'digital humanities'; this returned forty-one hits. We also searched for 'digital scholar' (zero results), 'digital history' (two results), 'digital classics' (two results), and 'digital edition' (eleven results). In a few cases the same project was indexed under two or more terms. Thus, the searches resulted in an initial set of forty-seven case studies. We then read each case study and identified the kinds of impact the case studies presented, and whether there was evidence for them. After an initial reading, it became apparent that, in some cases, the digital resource was either a very minor element of the whole project, or that the impact for it was either not claimed or not evidenced. This then left us with a set of forty-two studies.

Both the panel's evaluation and our reading of these case studies relied upon qualitative judgement because, while text-mining and statistical methods can show that the word 'museum' is present in a certain number of cases, we cannot tell how profound an effect, if any, the impact claimed on that museum or its visitors might be. Thus, we present findings in qualitative terms because we cannot know what judgements the panels themselves made, nor can we be sure that another reader looking at the same case studies would agree with every judgement we make.

All the case studies provided evidence for the use of their resource, in some cases on a very impressive scale. For example, the Diogenes software,[36] used to analyse classical texts, recorded 91,011 downloads, while the Old Bailey Online project[37] has had five million visits from 213 countries since 2003. In some instances, use and dissemination were confused with evidence of impact — a widespread issue in humanities subjects.[38] Numerous downloads of digital resources do not, of course, prove that users benefitted. However, all but four of the digital case studies did offer evidence of wide-ranging, genuine impact. Compared to the situation on which LAIRAH reported in 2005, where very few resource creators had any evidence of whether and how their resource was being used, this has been a huge step forward. It is also significant

---

36 *Diogenes*, https://community.dur.ac.uk/p.j.heslin/Software/Diogenes
37 *Old Bailey Online*, www.oldbaileyonline.org
38 REF, *Research Excellence Framework 2014: Overview Report by Main Panel D and Sub-Panels 27 to 36* (London: REF UK, 2015).

given that even in 2013 Maron et al. found that few resource creators had any contact with users, or collected data about use.[39]

In some ways, this apparent contradiction is explicable. Entering a research project of any kind, whether digital or not, as an REF impact case study was a highly selective process. The resources universities chose as case studies are likely to have been successful, and managed by dedicated PIs (principal investigators) and research teams who were likely to keep usage statistics. Once identified, case study projects had to collect further evidence of impact proactively. Nevertheless, the inclusion of the impact measurement in the REF appears to have produced an incentive for academics to keep information about how their research is used. Happily for DH, evidence of this is often easier to collect for digital resources than for analogue resources. In this sense at least, the impact measurement is, as Tanner argues, good for DH.[40]

Commercial Impact

We found several cases of commercial impact: DH's history of research in linguistic analysis resulted in the adoption of tools, algorithms, and resources outside academia. The GATE system,[41] developed by the University of Sheffield, has had a profound effect on commercial practices in natural language processing, as has the SCOTS corpus[42] from Glasgow on lexicography and the preparation of commercial teaching materials for English language. Software functionality for morphological analysis from the Diogenes system of Durham University was also used as part of a commercial publishing product: the Thesaurus Linguae Graecae.[43] Two spin-off companies were formed, both of which focused on digital imaging: Oxford Multi Spectral Ltd (University of Oxford) and Scriptura Ltd (http://scriptura.co.uk) (University of Sheffield). This is a relatively common practice in the sciences, but highly unusual for the humanities.

---

39   Maron, Yun, and Pickle, 'Sustaining our Digital Future'.
40   Simon Tanner, '3 Reasons Why REF2014 Was Good for Digital Humanities Scholars', *When the Data Hits the Fan!* (2 February 2015), http://simon-tanner.blogspot.co.uk/2015/02/3-reasons-ref2014-was-good-for-digital.html
41   *Gate: General Architecture for Text Engineering*, https://gate.ac.uk
42   *Scottish Corpus of Texts & Speech*, https://scottishcorpus.ac.uk
43   *Thesaurus Linguae Graecae*, http://stephanus.tlg.uci.edu/

Several projects gave rise to collaborations with the software industry, such as the University of Leeds' work on the *Cologne Edition* of Heinrich Böll, whose technical collaboration with software engineers at Pagina Ltd resulted in new software and platforms for large-scale critical editions. Perhaps the most unusual commercial relationship was the University of Westminster's collaboration with LEGO via digital community interaction and creativity.[44]

Media and Performance

Although broadcast media was most commonly used as a dissemination tool, we found several cases where digital projects had collaborated with the media to produce a genuine impact. Westminster researchers worked with the BBC and S4C to develop a virtual world for children called Adventure Rock. This research helped both companies reconsider their presentation of interactive experiences for children.[45] The complex nature of storytelling used by the Re-imagining the Literary Essay for the Digital Age (RILEDA) project at Brunel University (which created the multi-media digital literary essay *Kafka's Wound*)[46] changed the archiving practices of a media organisation (the London Review of Books)[47] and even gave rise to new forms of public performance, both live and recorded.

Work at the Centre for Robert Burns Studies,[48] including a digital edition,[49] has made numerous contributions to the Scottish cultural scene. In 2009, the project commissioned a new musical composition by Scottish composer James Macmillan, which was performed live, to mark

---

44 David Gauntlett, *Cultures of Creativity: Nurturing Creative Mindsets Across Cultures*, ed. by Bo Stjerne Thomsen (Billund: LEGO Foundation, 2013); David Gauntlett et al., *Defining Systematic Creativity in the Digital Realm* (Billund: LEGO Foundation, 2010).

45 David Gauntlett, 'Enabling and Constraining Creativity and Collaboration: Some Reflections after Adventure Rock', in *Content Cultures: Transformations of User Generated Content in Public Service Broadcasting*, ed. by Helen Thornham and Simon Popple (London: I. B. Tauris, 2013), pp. 161–80, https://doi.org/10.5040/9780755694426.ch-009

46 Will Self, *Kafka's Wound, a digital essay*, https://thespace.lrb.co.uk/

47 *London Review of Books*, www.lrb.co.uk

48 Centre for Robert Burns Studies, http://www.gla.ac.uk/schools/critical/research/researchcentresandnetworks/robertburnsstudies

49 *Editing Robert Burns for the 21st Century: An AHRC-Funded Project to Produce a Multi-Volume Edition of the Works of Robert Burns*, http://burnsc21.glasgow.ac.uk/

the 250th anniversary of Burns' birth. They also co-organised a successful world record attempt to perform Burns' *Auld Lang Syne* simultaneously in forty-one languages, which was recorded on YouTube.⁵⁰

University of Sussex and University of Cambridge's Newton Project,⁵¹ which provides an open access, online scholarly edition of Sir Isaac Newton's complete writings, inspired the play *Let Newton Be!*, along with other television and radio programmes, including BBC Radio 4's *In Our Time*, the BBC 4 series *The Beauty of Diagrams*, and BBC 2's *Isaac Newton: The Last Magician*.⁵² The University of Sheffield's *Old Bailey Online*,⁵³ a database of the records of criminal cases at the Old Bailey between 1674 and 1913, provided material for the BBC series *Tales from the Old Bailey*⁵⁴ and *Garrow's Law*.⁵⁵

Knowledge generated by digital projects and the use of digital linguistic analysis benefitted theatre companies, such as Shakespeare's Globe (University of Strathclyde's digital linguistic analysis as a rehearsal tool project at Shakespeare's Globe Theatre), the Royal Shakespeare Company (University of Birmingham's Debating Shakespeare in the Olympic Year Research), and King's College London's research project Out of the Wings: The Research and Practice of Spanish American Theatre in Translation benefited multiple theatres including the RSC, Silver Lining Theatre Company, CASA Festival, and the Royal Academy of Dramatic Art (RADA).

Cultural Heritage

Several projects fostered public engagement with cultural resources or the GLAM sector. We have written at greater length elsewhere about

---

50   Daily Record, 'Auld Lang Syne Record Set', *Youtube,* 1 December 2009, https://www.youtube.com/watch?v=9mb9ZwB_-xY&noredirect=1
51   *Newton Project*, www.newtonproject.sussex.ac.uk. The project team is currently based at the Faculty of History, University of Oxford.
52   'The Laws of Motion', *In Our Time*, BBC Radio 4, 3 April 2008, http://www.bbc.co.uk/programmes/b009mvj0; *The Beauty of Diagrams*, BBC 4, November–December 2010, http://www.bbc.co.uk/programmes/b00w5675
53   *Old Bailey Online*, www.oldbaileyonline.org
54   *Tales from the Old Bailey*, BBC 2, March–May 2013, http://www.bbc.co.uk/programmes/b01rdp8t
55   *Garrow's Law*, BBC 1, November 2009–February 2012, http://www.bbc.co.uk/programmes/b00w5c2w

our work on the QRator and Social Interpretation (SI) projects, which used digital resources to facilitate engagement with museums and were both the subject of case studies.[56] These projects were always designed to capture impact and evaluate the nature of benefit and change in visitor behaviour as part of the research projects and not for the sake of REF. However, it meant that we could provide evidence of impact in a way that few other digital projects were able to do.

Other innovative methods of engaging the public with cultural resources using digital methods were discussed in the case studies of crowd-sourced transcription projects. These included the groundbreaking Transcribe Bentham project,[57] and two projects from Oxford: the Oxyrhynchus Online,[58] and Ancient Lives,[59] which together made the Oxyrhynchus papyri available to the public using a web interface and crowdsourcing techniques. The facility to collect detailed evidence of the impact of the Transcribe Bentham project was built into the original research design, and has been published in greater detail than the case study word limit would allow.[60]

The London French project, from the University of Westminster, resulted in the creation of a community digital archive in collaboration with the British Library. This benefitted the French community, as well as information professionals, through the sharing of experiences and the dissemination of knowledge, and through the connections made between contemporary and historical lives. As a result of King's College London's Strandlines,[61] members of the local community

---

56 Claire Bailey-Ross et al., 'Engaging the Museum Space: Mobilizing Visitor Engagement with Digital Content Creation', *Digital Scholarship in the Humanities*, 32.4 (2016), 689–708, https://doi.org/10.1093/llc/fqw041; Claire Ross, Melissa Terras, and Carolyn Royston, 'Visitors, Digital Innovation and a Squander Bug: Reflections on Digital R&D for Audience Engagement and Institutional Impact', in *Museums and the Web 2013*, ed. by N. Proctor and R. Cherry (Silver Spring, MD: Museums and the Web, 2013); Mark Carnall, Jack Ashby, and Claire Ross, 'Natural History Museums as Provocateurs for Dialogue and Debate', *Museum Management and Curatorship*, 28.1 (2013), 37–41, https://doi.org/10.1080/09647775.2012.754630
57 *Transcribe Bentham*, www.transcribe-bentham.da.ulcc.ac.uk
58 'Oxyrhynchus Online', *Papyrology at Oxford*, www.papyrology.ox.ac.uk/POxy
59 *Ancient Lives*, www.ancientlives.org
60 Tim Causer and Valerie Wallace, 'Building a Volunteer Community: Results and Findings from Transcribe Bentham', *Digital Humanities Quarterly*, 6.2 (2012), http://www.digitalhumanities.org/dhq/vol/6/2/000125/000125.html
61 *Strandlines*, https://www.strandlines.london/

were able to interact in a digital public space with local artists, cultural practitioners, and creative industries to explore the meaning of place, discover the histories of their community, and exchange experiences. Research on a digital edition of the medieval Vernon Manuscript (Bodleian Library MS. Eng. poet. a. 1), written in the West Midlands' dialect, led to several public events in collaboration with some of Birmingham's libraries and museums.[62] This enhanced the understanding of the history and culture of the West Midlands and its contemporary dialect.

Several projects also benefited school-aged children and their teachers. Digital resources created by the University of Reading's Ure Museum of Greek Archaeology were used by school children at an animation workshop. The Ulster Poetry Project[63] developed an online library that has assisted in the development of teaching and learning materials about Ulster-Scots literature. Research on the eighteenth-century novelist Laurence Sterne at Northumbria University created a digital learning package for teachers to use when primary school children visit local heritage properties.[64] The *Candide* app, from the University of Oxford, is being used by secondary-aged students of Voltaire in French schools.[65]

As we have discussed above, projects such as these demonstrate that the impact of digital resources cannot always be categorised as academic-, community-, or GLAM-based. Indeed, such collaboration is vital to the success of many digital projects. We found numerous references to collaboration with the GLAM sector, including museums, galleries and libraries, and heritage sites, such as Norwich Cathedral, whose glass collection was made available digitally by the University of East Anglia's Norfolk Medieval Stained Glass Project.[66]

---

62 *Vernon Manuscript Project*, www.birmingham.ac.uk/vernonmanuscript
63 *Ulster Poetry Project*, arts.ulster.ac.uk/ulsterpoetry
64 'Learning Pack', *Dear Sterne*, http://dearsterne.blogspot.co.uk/p/learning-pack.html
65 *Lecture numérique: application 'Candide, edition enrichie'*, http://www.ac-grenoble.fr/mission-tice/Delegation_academique_au_numerique/Lecture_numerique_%3A_%22Candide%22.html
66 *Corpus Vitrearum Medii Aevi* (Medieval Stained Glass in Great Britain), www.cvma.ac.uk

Policy Impact

Perhaps more surprisingly, DH has also had an impact on public policy. The Clergy of the Church of England database (CCEd) 1540–1835 (University of Kent) resulted in changes in the ministry and practice of the Church of England;[67] while analysis of the language of 1641 depositions (a project by the University of Aberdeen) was used to facilitate public debate and political policy discussions about modern sectarianism in Northern Ireland.[68] The Freshwater Information Management project at King's College London has been used in environmental policy making, as well as to provide information to farmers and the public about water quality.[69] Material from Google Ancient Places (GAP) — an Open University project using GIS (geographic information systems) technology to map the ancient world — was used as part of the HathiTrust legal case in the USA, during which the right to fair access to digital educational materials was established.[70]

Limitations of the REF Case Studies

The REF case studies provide compelling evidence that DH has an impact beyond the predictable areas of the information professions and cultural heritage. However, there are limitations to the use of such material. The most obvious of these is that although REF criteria specify that impact should be judged on geographic reach, the exercise is not intended to benchmark impact in an international context. Although REF panels included members from the user community, digital resources created purely by the GLAM sector and commercial organisations without the input of academics, were excluded from the exercise. Information from the REF can be used to extrapolate the impact that the resources created outside the UK higher education sector might have, but there is no evidence base to test this in any meaningful way.

---

67 *Clergy of the Church of England Database*, http://theclergydatabase.org.uk/
68 *1641 Depositions Project*, http://www.abdn.ac.uk/1641-depositions/
69 Mark Hedges, Mike Haft, and Gareth Knight, 'FISHNet: Encouraging Data Sharing and Reuse in the Freshwater Science Community', *Journal of Digital Information*, 13.1 (2012).
70 *HathiTrust Opinion*, 2012, 11 CV 6351, p. 13, http://www.scribd.com/doc/109647049/HathiTrust-Opinion

It is also important to remember that the REF case studies were selected by universities and not randomly sampled. There was also no requirement to enter case studies that included digital tools or resources. This means that the case studies represent the strongest examples of the genre that could be found in any given university: cases where the impact of digital projects were difficult to prove were therefore not entered or evaluated. Thus, it is hardly surprising that the impact of such successful and high-profile projects was significant. We cannot, however, extrapolate from this that all digital resources must therefore have an impact: it is possible that most of them do not, or that it is only the most outstanding that do. We cannot ascertain what the ratio of outstanding, impactful projects to the average digital resource might be. The only way to test this would be to select digital resources at random from a list of funded projects, or from those archived in a repository, and then judge their impact accordingly.

This also leads to another limitation. The case studies were constructed and written by the universities themselves, who were responsible for collecting evidence of change or benefit, and for writing the narrative of the case study. However, such a procedure is naturally open to bias. Universities wanted to present their work in the best possible light and therefore selected evidence accordingly, perhaps disregarding indicators that were not as positive. A more objective method, whereby impact was judged by independent researchers against an agreed set of criteria, might reach different conclusions. However, doing this would be expensive and time intensive, and there is no evidence that there is any demand from funders, government, or the academics themselves, to carry out such an exercise.

Finally, while we are able to show that digital resources have an impact, so, it seems, does most research. In REF 2014, eighty-four percent of the impact section was judged to be four-star or three-star (eighty-one percent in panel D, which covered arts and humanities and digital resources). Thus, simply achieving impact for any research cannot be seen as exceptional, or even especially impressive.

Conclusions

The REF results demonstrate that DH can have an impact on numerous sectors, with some resources benefitting multiple sectors at a time. The case studies provide evidence of impact on cultural heritage, theatrical performance, the media, industry, schools, religious organisations, community groups, public policy, and the interested public. This is very welcome indeed. While such results are helpful in terms of advocacy for digital humanities, they are, nonetheless, of limited use to the creators of such resources themselves, if compared, for example, to Tanner's model. REF panels provided brief, general summaries of each unit of assessment, which sometimes contained comments on especially impressive impact cases. However, no detailed feedback was given, thus it is difficult for resource creators to know what was judged to be especially effective, or what might be improved. Tanner's model would probably have provided a more rigorous evaluation of the characteristics of such projects, but the time and funding required to undertake such a procedure may mean that, in an environment where resources are scarce, such protocols are relatively rarely used.

This recognition of the broad impact of DH is very heartening. The REF may be a positive force in bringing complex questions about the sustainability of digital resources to the fore. REF regulations allow the possibility of research having an impact up to twenty years after publication; and feedback on the 2014 exercise suggests this may still be too short a period, even in science and medicine. If we want digital resources to be able to have an impact for future REFs or other such exercises, they will still need to be accessible and functional beyond such a period — at the least. This is a significant challenge, given that, at present, the UK's Arts and Humanities Research Council only requires resource creators to ensure the availability of a resource for three years after the end of its funding period. It means that universities will need to think about how to plan for and fund the life of a digital resource for longer periods after the funding has ended. This entails not only making it available, but also keeping it updated, so that users feel confident in using it. By definition, if the functionality degrades, or the interface seems uninviting, and, as a result, use decreases, then evidence for longer-term impact will be harder to collect. This becomes even more complex in cases where a digital resource is a collaboration with, or even hosted by, a cultural heritage organisation, over whose sustainability policies universities do not have any control. But, of course, this only applies to resources hosted in the UK, there are no such levers elsewhere.

In this environment DH must, therefore, argue strongly for the impact of what it does so that in future its resources still exist to do so. As Nancy Maron and Sarah Pickle argue, DH is in an ideal position to demonstrate its impact.[71] DH resources are attractive and accessible to the public in a way that a dataset of scientific data simply cannot be. Not only have we built our resources so that they can be shared, but we can demonstrate that the public has been doing so, and indeed contributing to the content and intellectual endeavour of some digital projects. Impact is, as we have shown, not easy to capture or measure, but the experience of the REF suggests that we can offer evidence for the benefit and change brought about by DH resources in many different sectors.

Bibliography

1641 Depositions Project, http://www.abdn.ac.uk/1641-depositions/

Ancient Lives, www.ancientlives.org

Bailey-Ross, Claire et al., 'Engaging the Museum Space: Mobilizing Visitor Engagement with Digital Content Creation', *Digital Scholarship in the Humanities*, 32.4 (2016), 689–708, https://doi.org/10.1093/llc/fqw041

Bakhshi, Hasan, and David Throsby, *Culture of Innovation. An Economic Analysis of Innovation in Arts and Cultural Organizations* (Nesta, London, 2010).

The Beauty of Diagrams, BBC 4, November–December 2010, http://www.bbc.co.uk/programmes/b00w5675

Blake Archive, www.blakearchive.org

Carnall, Mark, Jack Ashby, and Claire Ross, 'Natural History Museums as Provocateurs for Dialogue and Debate', *Museum Management and Curatorship*, 28.1 (2013), 37–41, https://doi.org/10.1080/09647775.2012.754630

---

71   Nancy L. Maron, and Sarah Pickle, *Sustaining the Digital Humanities Host Institution Support beyond the Start-Up Phase* (New York: Ithaka S+R, 2014), pp. 15–16, https://doi.org/10.18665/sr.22548

Causer, Tim, and Valerie Wallace, 'Building a Volunteer Community: Results and Findings from Transcribe Bentham', *Digital Humanities Quarterly*, 6.2 (2012), http://www.digitalhumanities.org/dhq/vol/6/2/000125/000125.html

Centre for Robert Burns Studies, http://www.gla.ac.uk/schools/critical/research/researchcentresandnetworks/robertburnsstudies

Clergy of the Church of England Database, http://theclergydatabase.org.uk/

Corpus Vitrearum Medii Aevi (Medieval Stained Glass in Great Britain), www.cvma.ac.uk

Culture Metrics: A Shared Approach to Measuring Quality, http://www.culturemetricsresearch.com/

Daily Record, 'Auld Lang Syne Record Set', *Youtube*, 1 December 2009, https://www.youtube.com/watch?v=9mb9ZwB_-xY&noredirect=1

Department for Culture, Media, and Sport, *Statistical Data Set: Museums and Galleries Monthly Visits*, DCMS, London, 2017.

Diogenes, https://community.dur.ac.uk/p.j.heslin/Software/Diogenes

Editing Robert Burns for the 21st Century: An AHRC-Funded Project to Produce a Multi-Volume Edition of the Works of Robert Burns, http://burnsc21.glasgow.ac.uk/

Garrow's Law, BBC 1, November 2009–February 2012, http://www.bbc.co.uk/programmes/b00w5c2w

Gate: General Architecture for Text Engineering, https://gate.ac.uk

Gauntlett, David, *Cultures of Creativity: Nurturing Creative Mindsets Across Cultures*, ed. by Bo Stjerne Thomsen (Billund: LEGO Foundation, 2013).

—— 'Enabling and Constraining Creativity and Collaboration: Some Reflections after Adventure Rock', in *Content Cultures: Transformations of User Generated Content in Public Service Broadcasting*, ed. by Helen Thornham and Simon Popple (London: I. B. Tauris, 2013), pp. 161–80, https://doi.org/10.5040/9780755694426.ch-009

Gauntlett, David et al., *Defining Systematic Creativity in the Digital Realm* (Billund: LEGO Foundation, 2010).

HathiTrust Opinion, 2012, 11 CV 6351, http://www.scribd.com/doc/109647049/HathiTrust-Opinion

Hedges, Mark, Mike Haft, and Gareth Knight, 'FISHNet: Encouraging Data Sharing and Reuse in the Freshwater Science Community', *Journal of Digital Information*, 13.1 (2012).

Hellqvist, Björn, 'Referencing in the Humanities and its Implications for Citation Analysis', *Journal of the American Society for Information Science and Technology*, 61.2 (2010), 310–18, https://doi.org/10.1002/asi.21256

Higher Education Funding Council of England (HEFCE), *The Nature, Scale, and Beneficiaries of Research Impact: An Initial Analysis of Research Excellence Framework (REF) 2014 Impact Case Studies* (London: King's College London, 2015).

Holden, J., and J. Baltà, *The Public Value of Culture: A Literature Review* (EENC Paper, Brussels, 2012).

Hooper-Greenhill, Eilean, 'Measuring Learning Outcomes in Museums, Archives and Libraries: The Learning Impact Research Project (LIRP)', *International Journal of Heritage Studies*, 10.2 (2004), 151–74, https://doi.org/10.1080/1352725041ized0001692877

Hughes, Lorna. M., et al., 'Assessing and Measuring Impact of a Digital Collection in the Humanities: An Analysis of the SPHERE (Stormont Parliamentary Hansards: Embedded in Research and Education) Project', *Digital Scholarship in the Humanities*, 30.2 (2015), 183–98, https://doi.org/10.1093/llc/fqt054

Jones, Molly Morgan and Jonathan Grant, 'Making the Grade: Methodologies for Assessing and Evidencing Research Impact', in *7 Essays on Impact. DESCRIBE Project Report for Jisc*, ed. by David Cope et al. (Exeter: University of Exeter, 2013), pp. 25–43.

Keaney, Emily, 'Public Value and the Arts: Literature Review', *Strategy* (2006), 1–49.

Kinghorn, Naomi and Ken Willis, 'Measuring Museum Visitor Preferences Towards Opportunities for Developing Social Capital: An Application of a Choice Experiment to the Discovery Museum', *International Journal of Heritage Studies*, 14.6 (2008), 555–72, https://doi.org/10.1080/13527250802503290

'The Laws of Motion', *In Our Time*, BBC Radio 4, 3 April 2008, http://www.bbc.co.uk/programmes/b009mvj0

Lecture numérique: application 'Candide, edition enrichie', http://www.ac-grenoble.fr/mission-tice/Delegation_academique_au_numerique/Lecture_numerique_%3A_%22Candide%22.html

'Learning Pack', *Dear Sterne*, http://dearsterne.blogspot.co.uk/p/learning-pack.html

London Review of Books, www.lrb.co.uk

Maliniak, Daniel, Ryan Powers, and Barbara F. Walter, 'The Gender Citation Gap in International Relations', *International Organization*, 67.4 (2013), 889–922, https://doi.org/10.1017/s0020818313000209

Marcella, Rita, Hayley Lockerbie, and Lyndsay Bloice, 'Beyond REF 2014: The Impact of Impact Assessment on the Future of Information Research', *Journal of Information Science*, 42.3 (2016), 369–85, https://doi.org/10.1177/0165551516636291

Marcella, Rita, 'The Effects of the Research Excellence Framework Research Impact Agenda on Early- and Mid-Career Researchers in Library and

Information Science', *Journal of Information Science*, 44.5 (2018), 608–18, https://doi.org/10.1177/0165551517724685

Marchionni, Paola, 'Why Are Users So Useful? User Engagement and the Experience of the JISC Digitisation Programme', *Ariadne* (30 October 2009), http://www.ariadne.ac.uk/issue/61/marchionni/

Maron, Nancy L., Jason Yun, and Sarah Pickle, 'Sustaining our Digital Future: Institutional Strategies for Digital Content', *Strategic Content Alliance*, Ithaka Case Studies in Sustainability (2013), https://sca.jiscinvolve.org/wp/files/2013/01/Sustaining-our-digital-future-FINAL-31.pdf

Maron, Nancy L. and Sarah Pickle, *Sustaining the Digital Humanities Host Institution Support beyond the Start-Up Phase* (New York: Ithaka S+R, 2014), https://doi.org/10.18665/sr.22548

Matarasso, François, *Use or Ornament? The Social Impact of Participation in the Arts* (Stroud: Comedia, 1997).

Myerscough, John, *The Economic Importance of the Arts in Britain* (London: Policy Studies Institute, 1988).

Newton Project, www.newtonprojet.sussex.ac.uk

Old Bailey Online, www.oldbaileyonline.org

'Oxyrhynchus Online', *Papyrology at Oxford*, www.papyrology.ox.ac.uk/POxy

REF, 'Assessment Criteria and Level Definitions', https://www.ref.ac.uk/2014/panels/assessmentcriteriaandleveldefinitions/

—— *Assessment Framework and Guidance on Submissions* (Bristol: REF UK, 2011), http://www.ref.ac.uk/2014/media/ref/content/pub/assessmentframeworkandguidanceonsubmissions/GOS%20including%20addendum.pdf

—— *Research Excellence Framework 2014: Overview Report by Main Panel D and Sub-Panels 27 to 36*, (London: REF UK, 2015).

Ross, Claire, Melissa Terras, and Carolyn Royston, 'Visitors, Digital Innovation and a Squander Bug: Reflections on Digital R&D for Audience Engagement and Institutional Impact', in *Museums and the Web 2013*, ed. by N. Proctor and R. Cherry (Silver Spring, MD: Museums and the Web, 2013).

Rossetti Archive, www.rossettiarchive.org

Self, Will, *Kafka's Wound, A Digital Essay*, https://thespace.lrb.co.uk/

Selwood, Sara, 'Making a Difference: The Cultural Impact of Museums. An Essay for NMDC' (2010), https://www.nationalmuseums.org.uk/media/documents/publications/cultural_impact_final.pdf

—— 'What Difference Do Museums Make? Producing Evidence on the Impact of Museums', *Critical Quarterly*, 44.4 (2002), 65–81, https://doi.org/10.1111/1467-8705.00457

Scottish Corpus of Texts & Speech, https://scottishcorpus.ac.uk

Showers, Ben, 'A Strategic Approach to the Understanding and Evaluation of Impact', in *Evaluating and Measuring the Value, Use and Impact of Digital Collections*, ed. by Lorna M. Hughes (London: Facet, 2012), pp. 63–72, https://doi.org/10.29085/9781856049085.006

Smithies, James, et al.,'Managing 100 Digital Humanities Projects: Digital Scholarship & Archiving in King's Digital Lab', *Digital Humanities Quarterly*, 13.1 (2019), http://www.digitalhumanities.org/dhq/vol/13/1/000411/000411.html

Strandlines, https://www.strandlines.london/

Tales from the Old Bailey, BBC 2, March-May 2013, http://www.bbc.co.uk/programmes/b01rdp8t

Tanner, Simon and Marilyn Deegan, *Inspiring Research, Inspiring Scholarship. The Value and Benefits of Digitised Resources for Learning, Teaching, Research and Enjoyment* (London: JISC, 2011).

Tanner, Simon, *Measuring the Impact of Digital Resources: The Balanced Value Impact Model* (London: King's College London, 2012).

—— 'The Value and Impact of Digitized Resources for Learning, Teaching, Research and Enjoyment', in *Evaluating and Measuring the Value, Use and Impact of Digital Collections*, ed. by L. M. Hughes (London: Facet, 2012), pp. 103–20, https://doi.org/10.29085/9781856049085.009

—— '3 Reasons Why REF2014 Was Good for Digital Humanities Scholars', *When the Data Hits the Fan!* (2 February 2015), http://simon-tanner.blogspot.co.uk/2015/02/3-reasons-ref2014-was-good-for-digital.html

Thesaurus Linguae Graecae, http://stephanus.tlg.uci.edu/

'TIDSR: Toolkit for the Impact of Digitised Scholarly Resources', *Oxford Internet Institute*, https://www.oii.ox.ac.uk/research/projects/tidsr/

Transcribe Bentham, www.transcribe-bentham.da.ulcc.ac.uk

Travers, Tony, *Museums and Galleries in Britain Economic, Social and Creative Impacts* (London: London School of Economics & Political Science, 2006).

Ulster Poetry Project, arts.ulster.ac.uk/ulsterpoetry

Vernon Manuscript Project, www.birmingham.ac.uk/vernonmanuscript

Warwick, Claire, 'Archive 360: The Walt Whitman Archive', *Archive Journal*, 1.1 (2011).

Warwick, Claire, et al., 'If You Build It Will They Come? The LAIRAH Study: Quantifying the Use of Online Resources in the Arts and Humanities through Statistical Analysis of User Log Data', *Literary and Linguist Computing*, 23.1 (2008), 85–102, https://doi.org/10.1093/llc/fqm045

Wavell, Caroline, et al., *Impact Evaluation of Museums, Archives and Libraries: Available Evidence Project* (Aberdeen: Robert Gordon University, 2002).

West, Jevin D., et al., 'The Role of Gender in Scholarly Authorship', ed. by Lilach Hadany, *PLOS ONE*, 8.7 (2013), e66212, https://doi.org/10.1371/journal.pone.0066212

Whitman Archive, www.whitmanarchive.org

Wilkinson, Claire, 'Evidencing Impact: A Case Study of UK Academic Perspectives on Evidencing Research Impact', *Studies in Higher Education*, 44.1 (2019), 72–85, https://doi.org/10.1080/03075079.2017.1339028

Wilsdon, James, et al., *The Metric Tide: Report of the Independent Review of the Role of Metrics in Research Assessment and Management* (HEFCE: London, 2015), https://doi.org/10.13140/RG.2.1.4929.1363

5. Violins in the Subway

Scarcity Correlations, Evaluative Cultures, and Disciplinary Authority in the Digital Humanities

Martin Paul Eve

In January 2007, in a busy metro station in Washington DC, a violinist began to play. Of the 1097 people who walked by this violinist, twenty-seven contributed a dollar or so and seven stopped to listen. At the end of the three-quarter-hour session playing Bach, the violinist had amassed $32.17. Only one of the thousand or so passers-by recognised the busker as the world-famous virtuoso Joshua Bell who had, a mere three nights before, played the same repertoire at Boston's Symphony Hall with good tickets going for $100 each. The violin on which Bell performed in the subway was worth $3.5m.[1]

Although $30 per hour was not a bad rate of remuneration in the economic climate of 2007, the clearer point that emerges from the Bell experiment — a stunt fronted by *The Washington Post* — is that we are not very good, as a species, at identifying quality without frames of reference. As the found objects of the readymade Modernist period taught us of art: context is everything. What, though, if the same phenomena applied to scholarship? How good are we at independently judging research work, devoid of its enframing apparatus? Can we judge the music (the research) outside of the concert hall (the journal or press)?

---

1   Gene Weingarten, 'Pearls Before Breakfast: Can One of the Nation's Great Musicians Cut through the Fog of a D.C. Rush Hour? Let's Find Out.', *The Washington Post* (23 September 2014), https://www.washingtonpost.com/lifestyle/magazine/pearls-before-breakfast-can-one-of-the-nations-great-musicians-cut-through-the-fog-of-a-dc-rush-hour-lets-find-out/2014/09/23/8a6d46da-4331-11e4-b47c-f5889e061e5f_story.html

In the digital humanities (DH), this crux of evaluation has been entirely evident for some time.[2] A digital historian undertaking graduate study, for instance, reported a threat that they received: 'you will never gain a PhD doing this work'.[3] Those working in digital literary studies are advised to publish traditional works alongside their digital projects, effectively doubling the labour required of their analogous analogue cohort.[4] The very promise of performed new media — i.e. research artefacts that can grow and live, and that are born in the digital space — seems to re-stoke debates around print/scholarly fixity and the evaluation of ephemeral objects.[5] Essentially, those working in the digital humanities are told, time and time again, that their work will not count. At the same time, traditional scholars often perceive the digital humanities as a 'hot topic' in which it is easy to gain academic employment and tenure, leading to their fear of being crowded out. Certainly, as far back as 2011, the director of the National Endowment for the Humanities' office for the digital humanities, Brett Bobley, joked that there was a fear that DH was a 'secret plan to replace human scholars with robots'.[6]

Whence this conflicting sentiment? How can we understand this double logic in which DH work is at once so powerful as to crowd out the traditional humanists, while at the same time so poorly understood as to need supplementation with traditional publication? How can it be seen as both a sure-fire path to tenure but also a 'risky thing', as Kathleen Fitzpatrick and Mark Sample put it, to conduct digital labour

---

2   See Susan Schreibman, Laura Mandell, and Stephen Olsen, 'Introduction', *Profession* (2011), 123–201 https://doi.org/10.1632/prof.2011.2011.1.123; and the associated special issue for just one example.

3   The source of this quotation asked to remain unattributed, except to note that it was spoken by a woman of incredible fire and credulity.

4   Sydni Dunn, 'Digital Humanists: If You Want Tenure, Do Double the Work', *Vitae, the Online Career Hub for Higher Ed* (5 January 2014), https://chroniclevitae.com/news/249-digital-humanists-if-you-want-tenure-do-double-the-work

5   Helen J. Burgess and Jeanne Hamming, 'New Media in the Academy: Labor and the Production of Knowledge in Scholarly Multimedia', *Digital Humanities Quarterly (DHQ)*, 5.3 (2011), http://www.digitalhumanities.org/dhq/vol/5/3/000102/000102.html; see also Adrian Johns, *The Nature of the Book* (Chicago, IL: The University of Chicago Press, 1998), for a famous reply to Elizabeth Eisenstein [Hudson, Nicholas. "Challenging Eisenstein: Recent Studies in Print Culture." *Eighteenth-Century Life* 26, no. 2 (June 1, 2002): 83–95].

6   Steve Kolowich, 'The Promise of Digital Humanities', *Inside Higher Ed* (28 September 2011), https://www.insidehighered.com/news/2011/09/28/promise-digital-humanities

in the humanities?[7] Further, in this chapter I also ask whether there is something about the broader climate at the moment in academia that fears the collapse of traditional gatekeeping mechanisms. This is linked to longer-term trends in the digital availability of scholarship and scientific publication but also to the ways in which the abundance of the digital space allows for the publication of a wider range of artefacts. For while it is possible for scholars to publish diverse types of digital artefacts, and for these to be of a high-quality, to understand the challenges of awarding credit in the contemporary age requires an understanding of the shortage of labour time for evaluation, and the necessity of frames in the evaluation of research work.

Judging Excellence and Academic Hiring and Tenure

In order to understand the broader contexts of the academy within which DH evaluation takes place, we must ask a few further questions: just how bad are we at judging whether academic work is excellent? What about within niche sub-fields? And can we tell if work closely related to our own is any good?

As with the commuters who ignored Joshua Bell in the subway, the answers are: we are very bad at judging excellence, even within niche sub-fields closely related to our own. As I have noted elsewhere, alongside many others, researchers are extremely poor at judging quality even within their own fields.[8] This works both in terms of false negatives and false positives. For instance, in the former category, Juan Miguel Campanario, Joshua S. Gans, and George B. Shepherd examined instances of Nobel-prize winning work being rejected by

---

7   Kathleen Fitzpatrick, 'Do "the Risky Thing" in Digital Humanities', *The Chronicle of Higher Education* (25 September 2011), http://www.chronicle.com/article/Do-the-Risky-Thing-in/129132/; Mark Sample, 'Tenure as a Risk-Taking Venture', *Journal of Digital Humanities*, 1.4 (2012), http://journalofdigitalhumanities.org/1-4/tenure-as-a-risk-taking-venture-by-mark-sample/

8   Samuel Moore et al., '"Excellence R Us": University Research and the Fetishisation of Excellence', *Palgrave Communications*, 3 (2017), https://doi.org/10.1057/palcomms.2016.105; Adam Eyre-Walker and Nina Stoletzki, 'The Assessment of Science: The Relative Merits of Post-Publication Review, the Impact Factor, and the Number of Citations', *PLOS Biol*, 11.10 (2013), e1001675, https://doi.org/10.1371/journal.pbio.1001675

top journals.⁹ Campanario and others also note that there are many originally rejected papers that go on to be among the most highly cited in their fields.¹⁰ This is unsurprising, since most rejected manuscripts are eventually published elsewhere.¹¹ Even more worryingly though, is that there are also instances of false positives. In 1982, Douglas P. Peters and Stephen J. Ceci re-submitted disguised papers to psychology journals that had already accepted the same works for publication. They found that only eight percent were detected as plagiarised but that ninety percent were rejected on methodological and other grounds by journals in which the material had already appeared.¹² It is unclear precisely how these studies translate into the humanities disciplines, but it would not be a radical hypothesis to suggest that there may be analogies.

However, despite the evidence from the above studies, most academics are usually unwilling to admit that they are unable to determine quality. Were they to do so, the entire peer review mechanism would need to be dismantled.¹³ Instead, there is another factor present in the understanding of the instruments through which quality is assessed in the academy: a shortage of evaluative labour.

---

9   Juan Miguel Campanario, 'Rejecting and Resisting Nobel Class Discoveries: Accounts by Nobel Laureates', *Scientometrics*, 81.2 (2009), 549–65, https://doi.org/10.1007/s11192-008-2141-5; Joshua S. Gans and George B. Shepherd, 'How Are the Mighty Fallen: Rejected Classic Articles by Leading Economists', *The Journal of Economic Perspectives*, 8.1 (1994), 165–79, https://doi.org/10.1257/jep.8.1.165

10 Juan Miguel Campanario, 'Consolation for the Scientist: Sometimes It Is Hard to Publish Papers that Are Later Highly-Cited', *Social Studies of Science*, 23.2 (1993), 342–62, https://doi.org/10.1177/030631293023002005; Juan Miguel Campanario, 'Have Referees Rejected Some of the Most-Cited Articles of All Times?', *Journal of the American Society for Information Science*, 47.4 (1996), 302–10, https://doi.org/10.1002/(SICI)1097-4571(199604)47:4%3C302::AID-ASI6%3E3.0.CO;2-0; Juan Miguel Campanario and Erika Acedo, 'Rejecting Highly Cited Papers: The Views of Scientists Who Encounter Resistance to their Discoveries from Other Scientists', *Journal of the American Society for Information Science and Technology*, 58.5 (2007), 734–43 https://doi.org/10.1002/asi.20556; Kyle Siler, Kirby Lee, and Lisa Bero, 'Measuring the Effectiveness of Scientific Gatekeeping', *Proceedings of the National Academy of Sciences*, 112.2 (2015), 360–65 https://doi.org/10.1073/pnas.1418218112

11 See Moore et al., '"Excellence R Us"'.

12 Douglas P. Peters and Stephen J. Ceci, 'Peer-Review Practices of Psychological Journals: The Fate of Published Articles, Submitted Again', *Behavioral and Brain Sciences*, 5.2 (1982), 187–95, https://doi.org/10.1017/S0140525X00011183

13 Despite my criticisms here, it is certainly the case that peer review may spot errors. Experts are able to question matters of fact and interpretation. They are just not good at judging the value and/or worth of work in the present.

With up to four-hundred applicants for a single academic job, hiring panels often resort to proxy measures to evaluate quality.[14] In other words, there is insufficient labour on search committees to read and evaluate the research work of four hundred candidates, despite the fact that hiring for permanent/tenured positions represents a potential investment of several million dollars over the life of a career. While the final shortlist of candidates may have their work read, others are often eliminated by recourse to the press/journal name in which they were published, or nebulous citation measures such as the impact factor (IF).[15] This is clearly poor academic practice that does not allow for the discrepancy between the container and its contents, and which has led to declarations such as the 'San Francisco Declaration on Research Assessment' (DORA), disavowing such techniques.[16] For it is certainly the case, for example, that top university presses can publish bad books but also that low-ranking journals can contain gems. Academic freedom should entail the ability to submit one's work wherever one wishes. However, such freedom is severely constrained by this mechanism of proxy evaluation that concentrates material rewards upon specific publication brands.

This mechanism of proxy evaluation can 'work' for hiring panels because publication brand correlates with scarcity, as do the applicant-to-position ratios in universities. That is to say, for example, that if it is believed that having two books with top university presses is probably achievable by around one in four hundred candidates, then the proxy works perfectly for the above hypothetical hiring scenario. In this way, publication and evaluation through scarcity proxy measures act as a symbolic economy. The currency of this economy is research artefacts, which can be traded, through hiring, promotion, and tenure panels, into a real-world material economy (jobs, pay, benefits, healthcare, pensions) for the select few.

---

14 See Martin Paul Eve, *Open Access and the Humanities: Contexts, Controversies and the Future* (Cambridge: Cambridge University Press, 2014), chapter 2, https://doi.org/10.1017/CBO9781316161012

15 Björn Brembs, Katherine Button, and Marcus Munafò, 'Deep Impact: Unintended Consequences of Journal Rank', *Frontiers in Human Neuroscience*, 7 (2013), 291, https://doi.org/10.3389/fnhum.2013.00291

16 'San Francisco Declaration on Research Assessment', *DORA*, https://sfdora.org/read/

The essence of this evaluative culture is one that uses the frame of publication to judge the quality of research, usually problematically centred around a presumed single/individual author.[17] It is the same type of frame that uses the concert hall to judge the violinist, and that lacks discriminatory power when that same violinist appears in the subway. While it may be true, as Kathleen Fitzpatrick suggests, that 'we must be willing to engage in the act of judgment ourselves', we must also acknowledge the difficulties we face in undertaking such acts.[18] Importantly, though, this culture is also one that confers value upon specific media. University presses, for instance, publish books and journals. When 'university presses' are, then, the way in which hiring, promotion, and tenure panels make their decisions, there is an implicit underlying constraint of the valid forms that may be framed for evaluation through such proxy measures. Further, there is the matter of the continued belief in the efficacy of peer review, despite evidence to the contrary, which is linked to a reinforcement of existing media types. For example, if the labour of peer review is itself a type of service practice on which academics are assessed, the motivation to review for a high-profile press — whose brand will once again help with career advancement — is stronger than the motivation to review for radical/new publishers. This then reinforces the types of media that those traditional press entities publish, since peer review must be attached to particular objects and media types. In other words, there is a strong circularity of incentives for both authorship and peer-reviewing practices that severely constrict change in the type of media through which academics are assessed.

Finally, it is also necessary to pay attention to disciplinarity as a constraining factor in the evaluative cultures of university hiring. Disciplinary segregation, as Samuel Weber has charted it, is a way of

---

17 See Bethany Nowviskie, 'Where Credit Is Due: Preconditions for the Evaluation of Collaborative Digital Scholarship', *Profession* (2011), 169–81 https://doi.org/10.1632/prof.2011.2011.1.169; and Bethany Nowviskie, 'Evaluating Collaborative Digital Scholarship (Or, Where Credit Is Due)', *Journal of Digital Humanities*, 1.4 (2012), http://journalofdigitalhumanities.org/1-4/evaluating-collaborative-digital-scholarship-by-bethany-nowviskie/. For more on the challenges of adapting peer review for collaborative evaluation.

18 Kathleen Fitzpatrick, 'Peer Review, Judgment, and Reading', *Profession* (2011), 196–201 (p. 201), https://doi.org/10.1632/prof.2011.2011.1.196

amplifying authority through the delineation of a sub-field over which one may preside. In other words:

> [i]n order for the authority of the professional to be recognized as autonomous, the 'field' of his 'competence' had to be defined as essentially self-contained [...] In general, the professional sought to isolate in order to control.

and

> [t]he university, as it developed in the latter half of the nineteenth century, became the institutional expression and articulation of the culture of professionalism.[19]

But, as the old advice for graduate students used to run in the UK, while English programmes churned out competent critical theorists, there were no critical theory departments; one had to be a literature scholar. Even within the realm of the digital humanities, though, this urge towards bounding and containment in the name of intellectual authority is a vocal debate.[20] That said, it is frequently recognised that the digital humanities is an interdisciplinary space, even when it is difficult to define this term.[21] So, while citing an unpopular figure in digital humanities circles, it remains true that 'being interdisciplinary is so very hard to do'.[22] Indeed, the tendency of interdisciplinary methods to rest upon a dominant home discipline, while legitimating themselves as being 'interdisciplinary' through reference to an exotic other discipline, is omnipresent. But the sites of authority through evaluation in universities are primarily grouped around traditional disciplinary categories that can feel threatened by digital incursions.

---

19 Samuel Weber, *Institution and Interpretation*, Cultural Memory in the Present (Stanford: Stanford University Press, 2001), pp. 27–33.
20 Lisa Spiro, '"This is Why We Fight": Defining the Values of the Digital Humanities', in *Debates in the Digital Humanities*, ed. by Matthew K. Gold (Minnesota: University of Minnesota Press, 2012), pp. 16–35, https://doi.org/10.5749/minnesota/9780816677948.003.0003
21 See Julie Thompson Klein, *Interdisciplining Digital Humanities: Boundary Work in an Emerging Field* (Ann Arbor, MI: University of Michigan Press, 2014), chapter 1, https://doi.org/10.3998/dh.12869322.0001.001
22 Stanley Fish, 'Being Interdisciplinary Is So Very Hard to Do', *Profession* (1989), 15–22.

The Diverse Media Ecology of Digital Humanities

There are, in the taxonomy that I have sketched above, three conjoined and self-reinforcing elements of the academic evaluative cultures around research:

- a desired scarcity correlation between the research artefact and the position;
- a frame for evaluation that denotes scarcity, that is media specific, and that saves evaluative and reviewer labour;
- a set of disciplinary norms and agreements about which frames (in point 2) best denote comparable scarcity (in point 1).

The digital humanities, or in some cases just the digital, pose threats to a number of these evaluative cultures.

To begin to unpick this, consider that digital dissemination in general is causing problems for the scarcity correlation. In previous eras, the scarcity correlation was obtained through material print scarcity. That is, before digital dissemination was possible, a limited page budget with comparatively high printing costs per page came together to enforce a condition of scarcity. The digital environment changes this. In the world of the digital the vast majority of costs are shunted into the costs required to produce the first copy, which is still far from negligible in the academic publishing space (labour functions and estates costs include: typesetting, copyediting, proofreading, platform maintenance, digital preservation, identifier assignment, report generation, accountancy, legal, property, and equipment), while the costs of producing subsequent copies become almost zero. By decreasing unit cost and also by moving different forms of labour onto authors, as Matthew G. Kirschenbaum has recently noted, the print scarcity that previously underpinned the scarcity correlation for quality begins to collapse.[23] As journal articles and books — the previous media of print scarcity — become digital in their production, their scarcity function, which was always an economic function, is degraded. This is a little like the dropping of the gold standard as a way to measure the value of currency. Except, in the

---

23 Matthew G. Kirschenbaum, *Track Changes: A Literary History of Word Processing* (Cambridge, MA: The Belknap Press of Harvard University Press, 2016), chapter 3.

case of academic hiring, the belief in the value of the artefact, decoupled from any non-imagined scarcity, does not seem sufficient to continue.[24] In this way, at the heart of the digital's possibilities of infinite near-zero-cost dissemination lies the antithesis to the scarcity that has been used as a hiring proxy until now.

The practices of those working specifically in some form of the digital humanities, though, pose a set of additional challenges not only for the scarcity correlation but also for the frames of evaluation. In common with other scientific disciplines, the rise of the need to disseminate diverse forms of quantitative and qualitative data, software/code, and interactive artefacts within DH breaks the conditions of scarcity in a very particular way. For it is not precisely that such artefacts (the 'project as basic unit' as Anne Burdick et al. put it) are not scarce.[25] Whether it is the Digital Library of the Caribbean or the Манускрипт project of Udmurt State University and Izhevsk State Technical University that is under discussion, these projects are often unique; the ultimate form of scarcity. The same could be said, of course, of conventional academic books and articles, which are supposed to be unique in their original contributions to knowledge. Yet, books and journal articles are treated as comparable units of currency, while data, code, and interactive exhibits, in their uniqueness, are usually treated as though they were incomparable. In this way, there is a belief in the comparability of artefacts and the way they can be reviewed based on the shared, or otherwise, characteristics of their media form (digital vs print).

This supposition of (in)comparability is predicated on the belief of the uniform (or otherwise) nature of peer review that is tied to media form. Although the peer review's gatekeeping process is usually kept hidden due to concerns about anonymity and the freedom to speak truth to power, the furore around *PLOS ONE*'s reduced threshold of evaluation for publication illustrates this anxiety. In the *PLOS ONE*

---

24 There are some challenges with the divide that I am here drawing between a digital abundance and a material scarcity. Since labour is itself scarce and tied to material economic scarcity, and since there is labour in publishing, there remains a real non-imagined scarcity even in the digital. For more on this, see Martin Paul Eve, 'Scarcity and Abundance', in *The Bloomsbury Handbook of Electronic Literature*, ed. by Joseph Tabbi (London: Bloomsbury, 2017), pp. 385–98, https://doi.org/10.5040/9781474230285.ch-022

25 Anne Burdick et al., *Digital Humanities* (Cambridge, MA: MIT Press, 2012), pp. 124–25.

model, work is appraised on its technical soundness rather than on its novelty, originality, or significance. This mode of peer review is designed to encourage replication studies and the publication of negative results, aspects that are also of interest to many data-driven sub-fields in DH, such as stylometry and sentiment analysis. However, as noble as its scientific purposes may be, *PLOS ONE*'s altered review model causes substantial problems for hiring and evaluative proxies. How should the name *PLOS ONE* be viewed alongside *Nature* or *Science*? In deliberately lowering its scarcity threshold in the name of good science, *PLOS ONE* asked the academic community to examine its own processes for evaluation. In making itself unique as a mega-journal with this threshold, the brand of the journal has been altered. Yes, *PLOS ONE* was, at the time, itself scarce as the only entity of its type, but the quality threshold was not determined as comparable with other outlets and so the scarcity function was eroded. The unit of currency became non-exchangeable.

A similar problem occurs in one-off DH projects. Uniquely scarce, of course; these artefacts contribute to the diverse media ecology of the digital humanities. Yet, their very uniqueness, while being scarce, is non comparable. That is, because they are one-offs, developing standards for comparability is a disproportionate activity in terms of labour time, that does not fulfil the second characteristic I outlined above. In other words, to evaluate the artefact, as itself, rather than through a proxy of presumed-uniform review, has no labour-saving function. This is why such unframed projects begin to cause anxiety among those who have come to rely on the proxies that they believe denote comparable scarcity.

This is, in part, why we have seen the emergence of documents such as the MLA's 'Guidelines for Evaluating Work in Digital Humanities and Digital Media' or the AHA's 'Guidelines for the Professional Evaluation of Digital Scholarship by Historians'.[26] However, these guiding

---

26   Modern Language Association of America, 'Guidelines for Evaluating Work in Digital Humanities and Digital Media', *Modern Language Association* (2012), https://www.mla.org/About-Us/Governance/Committees/Committee-Listings/Professional-Issues/Committee-on-Information-Technology/Guidelines-for-Evaluating-Work-in-Digital-Humanities-and-Digital-Media; American Historical Association, 'Guidelines for the Professional Evaluation of Digital Scholarship by Historians', *American Historical Association* (2015), https://www.historians.org/teaching-and-learning/digital-history-resources/evaluation-of-digital-scholarship-in-history/guidelines-for-the-professional-evaluation-of-digital-scholarship-by-historians

documents often struggle to fulfil the 'needs' of hiring committees. That is, in asking for respect for medium specificity, alongside the requirement for the engagement of qualified reviewers — or, as Sheila Cavanagh puts it, by asking for consideration of the 'complicated factors in the world of digital scholarship needing attention' — such guidelines do not alleviate the labour shortage of the search panels nor do they provide a uniform comparability mechanism for scarcity.[27] Although these go unacknowledged, since most panel members do not wish to admit that they need recourse to such proxies, the continued fetishisation of print (for its scarcity) and the desire for hidden, yet claimed, uniform and comparable media-constraining gatekeeping practices, all highlight why it remains difficult for the proliferation of new digital artefacts to be easily integrated within conventional hiring mechanisms.[28]

Strategies for Changing Cultures: Disciplinary Segregation, Print Simulation, and Direct Economics

The diverse media ecology of DH poses a threat to the first two areas in which hiring panels and accreditation mechanisms operate: in the abundance of its artefacts, the digital disrupts scarcity, while in the uniqueness of its outputs, it defies the comparability of proxy frames. The final area in which DH causes anxiety is in its inter-/multi-/transdisciplinary nature. The challenge that DH creates in this final space is one of both evaluation and authority. In the first case, conventional hiring panels often struggle to evaluate part of a DH project; that is, the digital part. A lack of statistical knowledge among members of a search committee can also cause trouble here for certain types of DH practice. The authority challenge that is posed here is an unseating from their thrones of those with insufficient digital knowledge to carry out evaluation. This is the same challenge that other fields, such as religious history, can face: to be hated by both theologians and historians. To have created a 'discipline' usually means that one understands the

---

27 Sheila Cavanagh, 'Living in a Digital World: Rethinking Peer Review, Collaboration, and Open Access', *Journal of Digital Humanities*, 1.4 (2012), https://doi.org/10.5038/2157-7129.2.1.14, http://journalofdigitalhumanities.org/1-4/living-in-a-digital-world-by-sheila-cavanagh/

28 I am aware that there are other good reasons to stick to print for long-form reading. However, in the assessment domain, it is the scarcity that is valued.

evaluative requirement within that space. The practices of DH, which can intrude upon any of the conventional humanities disciplines, are challenging to those at the top of the pyramid since they suddenly find that they are not masters of their own kingdom. The work purports to be in a subject area that is recognisable to them but they know neither how to evaluate it nor how to test the research for relative soundness. When a discipline cannot evaluate work that purports to be within its own subject area, it faces a crisis. Hence why DH poses such a threat. Max Planck once famously put it that science advances 'one funeral a time'. Since disciplines are self-reinforcing spaces, though, it is not even clear that this is the case; value systems are absorbed and internalised by those who travel through the academic ranks.[29]

One of the strategies for avoiding this interdisciplinary threat has been to establish and strengthen specific DH departments. At the time of writing the most recent example of this was at King's College London where the department of Digital Humanities advertised for eight permanent, full-time posts (tenured equivalents) ranging from lecturers up to full professors. Likewise, the School of Advanced Study at the University of London is seeking a candidate to lead a new national centre for digital practice. By demarcating the space of expertise to a specifically digital domain it is possible to pursue digital practices and to hire staff members in ways that do not appear to compromise disciplinary expertise or authority. On the other hand, this also leads to a potentially problematic 'siloization' of digital expertise and the merely static reproduction of other disciplinary norms on which it is often the purpose of DH to intrude. For example: what use are authorship attribution technologies if nobody who defines themselves as a traditional literary scholar pays any attention? What is the point of spatio-temporal mapping approaches and GIS techniques if they cannot be used to inform other disciplinary cultures? From a research point of view, the banishment of DH to its own departmental area is a problematic move.

That said, DH as a departmental space makes sense from the economic perspective of teaching. Such programmes, which can often

---

29   Pierre Azoulay, Christian Fons-Rosen, and Joshua S. Graff Zivin, *Does Science Advance One Funeral at a Time?* (Cambridge, MA: National Bureau of Economic Research, 2015), https://doi.org/10.3386/w21788

promise transferable practical skills training and general computational thinking, recruit well; although they have also come under fire for apparently selling out and instrumentalising the humanities.[30] The general difficulty, though, is in the intellectual breadth covered by the single seemingly simple word: humanities. Some humanists have more in common with mathematics than with literature, while others are more akin to social sciences. In the disciplinary segregation of DH then, at the same time as providing for a broader perspective and harnessing the benefits of a wide set of views that transcend any single discipline, problems of an incoherent intellectual space can emerge. It is unclear, though, at least to me, whether computational approaches are enough to bind together such otherwise disparate fields of practice in perpetuity. For the sake of binding together these fields into a space of intellectual authority, we may see a set of changes — positive or negative — around disciplinary coherence.

A similar separation of DH is evident in the proliferation of new publishing venues for the field.[31] That many of these are still journals (the *Journal of Digital Humanities*, *Digital Humanities Quarterly*, and *Digital Scholarship in the Humanities*, are just three examples) speaks to a deep understanding among many digital humanists of the challenges of evaluative framing and media outlined above. Even though hiring panels could delegate evaluative authority to a DH community that somehow gatekept projects, by adhering to the understood mediaform of the journal article, research outputs become an exchangeable currency in diverse disciplinary settings. Similarly, book chapters are a recognisable form that play into the long history of the codex, but that are, in digital form, mostly a simulation of print. Such a simulation is effective since it appears to be a simulation of the form of material scarcity that was previously inherent within print. In other words, even while the greatest costs continue to inhere in selectivity, print simulation

---

30   Daniel Allington, Sarah Brouillette, and David Golumbia, 'Neoliberal Tools (and Archives): A Political History of Digital Humanities', *Los Angeles Review of Books* (1 May 2016), https://lareviewofbooks.org/article/neoliberal-tools-archives-political-history-digital-humanities/

31   Without veering too far into the 'defining DH' genre, see Alan Liu, 'Is Digital Humanities a Field? — An Answer from the Point of View of Language', *Alan Liu* (6 March 2013), http://liu.english.ucsb.edu/is-digital-humanities-a-field-an-answer-from-the-point-of-view-of-language/ for more on the use of the term 'field'.

is maintained so that the illusion of scarcity economics can be preserved within our systems of evaluation.

This notion of the simulation of other forms that DH has had to adopt is profitable. Indeed, many scientific disciplines also feel this pain of separation between the research outputs they produce and the work they conduct. This is why the recent practice of data sharing has simultaneously become both a welcome activity and a contentious one. Billing the sharing of data as better for replication and verification is an easy argument to make. Without it, journal articles are just descriptions of work without the underlying work itself: a print simulation of non-print activities. On the other hand, very few scientists would consider submitting a dataset to any evaluation exercise as the work itself.[32] The same goes for software and toolsets in the digital humanities; as Susan Schreibman and Ann M. Hanlon found, there is a 'relationship to [the] scholarship' of software in which many creators feel their work to be a scholarly activity, even while claiming more distant publication benefits.[33]

The final frame to which DH can, and does, resort is to bypass the symbolic economy entirely and move to hard currency: cash. DH is a relatively successful field in the space of research grants. Sheila Brennan addresses this: it is possible to 'let the grant do the talking'; that is, at once to allow the fact that DH attracts money to be itself a criterion for evaluation, but also to use the accountability and documentation practices to produce an archive of creditable narrative statements around a project.[34] Given that all systems of evaluations are economies, the cry of 'show me the money' can ring loudly. Yet, this is not likely to endear DH to traditional humanists, and it is not clear that DH will itself be spared the axe when the time comes.

In this chapter, I have explored how and why various systems from peer review and aggregation, to 'container-level' evaluation,

---

32 There have also been concerns raised about so-called 'research parasites' feeding off the data of others, although this seems like a logical and sensible practice to me. See Dan L. Longo and Jeffrey M. Drazen, 'Data Sharing', *New England Journal of Medicine*, 374.3 (2016), 276–77, https://doi.org/10.1056/NEJMe1516564

33 Susan Schreibman and Ann M. Hanlon, 'Determining Value for Digital Humanities Tools: Report on a Survey of Tool Developers', *DHQ: Digital Humanities Quarterly*, 4.2 (2010), http://www.digitalhumanities.org/dhq/vol/4/2/000083/000083.html

34 Sheila Brennan, 'Let the Grant Do the Talking', *Journal of Digital Humanities*, 1.4 (2012), http://journalofdigitalhumanities.org/1-4/let-the-grant-do-the-talking-by-sheila-brennan/

remain extremely limited, and yet are still in use. I have also made the case that all such systems of evaluation are economic in character. In turn, I have examined how the digital humanities field poses a set of challenges to the three principles of academic evaluation that I have outlined. It seems to me that it is very difficult to change the academic contexts of evaluation; they are complex social constructs (which is not to say that they do not have definitive real-world effects), not fixable technical realities. This gives a set of rationales for why DH continues to adopt publication practices that can be brought into harmony with such demands for substitution and exchange. While Lisa Samuels and Jerome J. McGann write of deformance, publication practice — for reasons of evaluation — remains in the realm of conformance, and will continue to remain there until we build our own disciplinary spaces.[35] These too, over time, will solidify their evaluative cultures and become unyielding to, and impenetrable by, new practices. In the meantime, listen for violins in the subway when next you ride.

Bibliography

Allington, Daniel, Sarah Brouillette, and David Golumbia, 'Neoliberal Tools (and Archives): A Political History of Digital Humanities', *Los Angeles Review of Books* (1 May 2016), https://lareviewofbooks.org/article/neoliberal-tools-archives-political-history-digital-humanities/

American Historical Association, 'Guidelines for the Professional Evaluation of Digital Scholarship by Historians', *American Historical Association* (2015), https://www.historians.org/teaching-and-learning/digital-history-resources/evaluation-of-digital-scholarship-in-history/guidelines-for-the-professional-evaluation-of-digital-scholarship-by-historians

Azoulay, Pierre, Christian Fons-Rosen, and Joshua S. Graff Zivin, *Does Science Advance One Funeral at a Time?* (Cambridge, MA: National Bureau of Economic Research, 2015), https://doi.org/10.3386/w21788

Brembs, Björn, Katherine Button, and Marcus Munafò, 'Deep Impact: Unintended Consequences of Journal Rank', *Frontiers in Human Neuroscience*, 7 (2013), 291, https://doi.org/10.3389/fnhum.2013.00291

Brennan, Sheila, 'Let the Grant Do the Talking', *Journal of Digital Humanities*, 1 (2012), http://journalofdigitalhumanities.org/1-4/let-the-grant-do-the-talking-by-sheila-brennan/

---

35 Lisa Samuels and Jerome J. McGann, 'Deformance and Interpretation', *New Literary History*, 30.1 (1999), 25–56 https://doi.org/10.1353/nlh.1999.0010

Burdick, Anne, et al., *Digital Humanities* (Cambridge, MA: MIT Press, 2012).

Burgess, Helen J., and Jeanne Hamming, 'New Media in the Academy: Labor and the Production of Knowledge in Scholarly Multimedia', *Digital Humanities Quarterly (DHQ)*, 5 (2011), http://www.digitalhumanities.org/dhq/vol/5/3/000102/000102.html

Campanario, Juan Miguel, 'Consolation for the Scientist: Sometimes It Is Hard to Publish Papers that Are Later Highly-Cited', *Social Studies of Science*, 23 (1993), 342–62, https://doi.org/10.1177/030631293023002005

—— 'Have Referees Rejected Some of the Most-Cited Articles of All Times?', *Journal of the American Society for Information Science*, 47 (1996), 302–10, https://doi.org/10.1002/(SICI)1097-4571(199604)47:4%3C302::AID-ASI6%3E3.0.CO;2-0

—— 'Rejecting and Resisting Nobel Class Discoveries: Accounts by Nobel Laureates', *Scientometrics*, 81 (2009), 549–65, https://doi.org/10.1007/s11192-008-2141-5

Campanario, Juan Miguel, and Erika Acedo, 'Rejecting Highly Cited Papers: The Views of Scientists Who Encounter Resistance to their Discoveries from Other Scientists', *Journal of the American Society for Information Science and Technology*, 58 (2007), 734–43, https://doi.org/10.1002/asi.20556

Cavanagh, Sheila, 'Living in a Digital World: Rethinking Peer Review, Collaboration, and Open Access', *Journal of Digital Humanities*, 1 (2012), https://doi.org/10.5038/2157-7129.2.1.14, http://journalofdigitalhumanities.org/1-4/living-in-a-digital-world-by-sheila-cavanagh/

Dunn, Sydni, 'Digital Humanists: If You Want Tenure, Do Double the Work', *Vitae, the Online Career Hub for Higher Ed* (5 January 2014), https://chroniclevitae.com/news/249-digital-humanists-if-you-want-tenure-do-double-the-work

Eve, Martin Paul, *Open Access and the Humanities: Contexts, Controversies and the Future* (Cambridge: Cambridge University Press, 2014), https://doi.org/10.1017/CBO9781316161012

—— 'Scarcity and Abundance', in *The Bloomsbury Handbook of Electronic Literature*, ed. by Joseph Tabbi (London: Bloomsbury, 2017), pp. 385–98, https://doi.org/10.5040/9781474230285.ch-022

Eyre-Walker, Adam, and Nina Stoletzki, 'The Assessment of Science: The Relative Merits of Post-Publication Review, the Impact Factor, and the Number of Citations', *PLOS Biol*, 11 (2013), e1001675, https://doi.org/10.1371/journal.pbio.1001675

Fish, Stanley, 'Being Interdisciplinary Is So Very Hard to Do', *Profession* (1989), 15–22.

Fitzpatrick, Kathleen, 'Do "the Risky Thing" in Digital Humanities', *The Chronicle of Higher Education* (25 September 2011), http://www.chronicle.com/article/Do-the-Risky-Thing-in/129132/

—— 'Peer Review, Judgment, and Reading', *Profession* (2011), 196–201, https://doi.org/10.1632/prof.2011.2011.1.196

Gans, Joshua S., and George B. Shepherd, 'How Are the Mighty Fallen: Rejected Classic Articles by Leading Economists', *The Journal of Economic Perspectives*, 8 (1994), 165–79, https://doi.org/10.1257/jep.8.1.165

Johns, Adrian, *The Nature of the Book* (Chicago, IL: The University of Chicago Press, 1998).

Kirschenbaum, Matthew G., *Track Changes: A Literary History of Word Processing* (Cambridge, MA: The Belknap Press of Harvard University Press, 2016).

Klein, Julie Thompson, *Interdisciplining Digital Humanities: Boundary Work in an Emerging Field* (Ann Arbor, MI: University of Michigan Press, 2014), https://doi.org/10.3998/dh.12869322.0001.001

Kolowich, Steve, 'The Promise of Digital Humanities', *Inside Higher Ed* (28 September 2011), https://www.insidehighered.com/news/2011/09/28/promise-digital-humanities

Liu, Alan, 'Is Digital Humanities a Field? — An Answer from the Point of View of Language', *Alan Liu* (6 March 2013), http://liu.english.ucsb.edu/is-digital-humanities-a-field-an-answer-from-the-point-of-view-of-language/

Longo, Dan L., and Jeffrey M. Drazen, 'Data Sharing', *New England Journal of Medicine*, 374 (2016), 276–77, https://doi.org/10.1056/NEJMe1516564

Modern Language Association of America, 'Guidelines for Evaluating Work in Digital Humanities and Digital Media', *Modern Language Association* (2012), https://www.mla.org/About-Us/Governance/Committees/Committee-Listings/Professional-Issues/Committee-on-Information-Technology/Guidelines-for-Evaluating-Work-in-Digital-Humanities-and-Digital-Media

Moore, Samuel, et al., '"Excellence R Us": University Research and the Fetishisation of Excellence', *Palgrave Communications*, 3 (2017), https://doi.org/10.1057/palcomms.2016.105

Nowviskie, Bethany, 'Evaluating Collaborative Digital Scholarship (Or, Where Credit Is Due)', *Journal of Digital Humanities*, 1 (2012), http://journalofdigitalhumanities.org/1-4/evaluating-collaborative-digital-scholarship-by-bethany-nowviskie/

—— 'Where Credit Is Due: Preconditions for the Evaluation of Collaborative Digital Scholarship', *Profession* (2011), 169–81, https://doi.org/10.1632/prof.2011.2011.1.169

Peters, Douglas P., and Stephen J. Ceci, 'Peer-Review Practices of Psychological Journals: The Fate of Published Articles, Submitted Again', *Behavioral and Brain Sciences*, 5 (1982), 187–95, https://doi.org/10.1017/S0140525X00011183

Sample, Mark, 'Tenure as a Risk-Taking Venture', *Journal of Digital Humanities*, 1 (2012), http://journalofdigitalhumanities.org/1-4/tenure-as-a-risk-taking-venture-by-mark-sample/

Samuels, Lisa, and Jerome J. McGann, 'Deformance and Interpretation', *New Literary History*, 30 (1999), 25–56, https://doi.org/10.1353/nlh.1999.0010

'San Francisco Declaration on Research Assessment', *DORA*, https://sfdora.org/read/

Schreibman, Susan, and Ann M. Hanlon, 'Determining Value for Digital Humanities Tools: Report on a Survey of Tool Developers', *DHQ: Digital Humanities Quarterly*, 4.2 (2010), http://www.digitalhumanities.org/dhq/vol/4/2/000083/000083.html

Schreibman, Susan, Laura Mandell, and Stephen Olsen, 'Introduction', *Profession* (2011), 123–201, https://doi.org/10.1632/prof.2011.2011.1.123

Siler, Kyle, Kirby Lee, and Lisa Bero, 'Measuring the Effectiveness of Scientific Gatekeeping', *Proceedings of the National Academy of Sciences*, 112 (2015), 360–65 https://doi.org/10.1073/pnas.1418218112

Spiro, Lisa, '"This is Why We Fight": Defining the Values of the Digital Humanities', in *Debates in the Digital Humanities*, ed. by Matthew K. Gold (Minnesota: University of Minnesota Press, 2012), pp. 16–35, https://doi.org/10.5749/minnesota/9780816677948.003.0003

Weber, Samuel, *Institution and Interpretation*, Cultural Memory in the Present (Stanford: Stanford University Press, 2001).

Weingarten, Gene, 'Pearls Before Breakfast: Can One of the Nation's Great Musicians Cut through the Fog of a D.C. Rush Hour? Let's Find Out.', *The Washington Post* (23 September 2014), https://www.washingtonpost.com/lifestyle/magazine/pearls-before-breakfast-can-one-of-the-nations-great-musicians-cut-through-the-fog-of-a-dc-rush-hour-lets-find-out/2014/09/23/8a6d46da-4331-11e4-b47c-f5889e061e5f_story.html

6. 'Black Boxes' and True Colour — A Rhetoric of Scholarly Code

Joris J. van Zundert, Smiljana Antonijević, and Tara L. Andrews

Introduction

Software pervades society. As Lev Manovich, Steven Jones, and David Berry have shown, there is hardly any form of contemporary data or information that has not been touched by digital means at some point during its creation.[1] The humanities, whose scholars study the data and information that is connected to social, historical, and cultural artefacts, are affected by a similar pervasiveness of software.[2] Programmers write software in a form of text known as source code: a series of instructions for how to perform a task, or a set of tasks, that the computer carries out. As software pervades the humanities, so its source code increasingly forms part of the makeup of the method and design in research projects in the humanities fields; this

---

1   Lev Manovich, *Software Takes Command: Extending the Language of New Media*, International Texts in Critical Media Aesthetics 5 (New York, NY: Bloomsbury Academic, 2013); Steven E. Jones, *The Emergence of the Digital Humanities* (New York, NY: Routledge, 2014); David M. Berry, *Critical Theory and the Digital*, Critical Theory and Contemporary Society (New York, NY: Bloomsbury Academic, 2014), https://doi.org/10.5040/9781501302114

2   Cf. Jones, *Emergence of the Digital Humanities*; Manovich, *Software Takes Command*.

is the particular focus of the emerging discipline, or methodology, or movement, known as digital humanities (DH).

As the expressions of a *technē* whose inner workings are opaque to most humanities scholars, code and codework[3] are all too often treated as invisible hands, which influence humanities research in ways that are neither transparent nor accounted for. The software used in research is treated as a 'black box' in the sense of information science — that is, it is expected to produce a certain output given a certain input — but, at the same time, it is often mistrusted precisely for this same lack of transparency. It is also often perceived as a mathematical — and thus value-neutral and socially inert — instrument; moreover, these two seemingly contradictory perceptions need not be mutually exclusive.

The lack of knowledge about what is actually taking place in these software 'black boxes' and about how they are made introduces serious problems for evaluation and trust in humanities research. If we cannot read code or see the workings of the software as it functions, we can experience it only in terms of its interface and its output, neither of which seem subject to our control. Yet, code is written by people, thus making it a social construct that embeds and expresses social and ideological beliefs of which it is — intentionally or not, directly or as a side effect — an agent.[4] Code is a more or less a withdrawn or even covert, but non-neutral, technology.[5] Therefore, when humanities scholars use software, they may unwittingly import certain methodological and epistemological assumptions inherent in that software into their research fields. Moreover, the invisibility and un-critiqued use of code in the humanities means that the scholarly quality and contribution of codework goes both uncredited and unaccounted for. To mitigate problems with academic evaluation and credit, a much greater insight into code and codework in the humanities

---

[3] We understand 'codework' to mean all the work involved in creating software source code that is more than just the act of writing the code. As we will explain further on, it encompasses many concrete and cognitive scholarly tasks. We use 'codework' as a broadly inclusive term, while we use 'coding' more narrowly as the act of writing source code.

[4] Tara McPherson, 'Why Are the Digital Humanities So White? Or Thinking the Histories of Race and Computation', in *Debates in the Digital Humanities*, ed. by Matthew K. Gold (Minneapolis: University of Minnesota Press, 2012), pp. 139–60, https://doi.org/10.5749/minnesota/9780816677948.003.0017, http://dhdebates.gc.cuny.edu/debates/text/29

[5] David M. Berry, *The Philosophy of Software: Code and Mediation in the Digital Age* (Basingstoke: Palgrave Macmillan, 2011).

is urgently required by those who engage in such evaluation; for instance, how coders approach their tasks, what decisions go into its production, and how code interacts with its environment. The purpose of this chapter is to provide some of that insight in the form of an ethnography of codework, wherein we observe the decisions that programmers make, and how they understand their own activities. This 'studying-up'[6] of people who hold epistemological and methodological power — in this case coding power — follows in the footsteps of ethnographies of technoscientific practice[7] and reflections on coding and tool development in DH.[8] Like other ethnographic studies, our small-scale exploration does not aspire to be fully representative of DH codework, but to initiate a debate about some still overlooked elements of this practice. We conclude this chapter with a discussion about our findings and several recommendations for how codework should be approached by programmers, scholars, and administrators in the humanities.

Background

Code can be understood as an argument in a way that is congruent with Alan Galey and Stan Ruecker's understanding of the epistemological status of graphical user interfaces as argument.[9] Code and codework

---

6 Laura Nader, 'Up the Anthropologist: Perspectives Gained from Studying Up', in *Reinventing Anthropology*, ed. by D. H. Hymes, Ann Arbor Paperbacks (University of Michigan Press, 1972), pp. 284–311, http://www.dourish.com/classes/readings/Nader-StudyingUp.pdf

7 E. Gabriella Coleman, *Coding Freedom: The Ethics and Aesthetics of Hacking* (Princeton (US), Woodstock (UK): Princeton University Press, 2013), http://gabriellacoleman.org/Coleman-Coding-Freedom.pdf; G. Coleman, *Hacker, Hoaxer, Whistleblower, Spy: The Many Faces of Anonymous* (London, New York: Verso, 2014); D. Forsythe, and D. J. Hess, *Studying Those Who Study Us: An Anthropologist in the World of Artificial Intelligence* (Stanford, CA: Stanford University Press, 2001).

8 Stephen Ramsay and Geoffrey Rockwell, 'Developing Things: Notes toward an Epistemology of Building in the Digital Humanities', in *Debates in the Digital Humanities*, ed. by Matthew K. Gold (Minneapolis: University of Minnesota Press, 2012), pp. 75–84, https://doi.org/10.5749/minnesota/9780816677948.003.0010, http://dhdebates.gc.cuny.edu/debates/text/11; Susan Schreibman and Ann M. Hanlon, 'Determining Value for Digital Humanities Tools: Report on a Survey of Tool Developers', *Digital Humanities Quarterly*, 4.2 (2010), http://digitalhumanities.org/dhq/vol/4/2/000083/000083.html; Nikolai Bezroukov, 'Open Source Software Development as a Special Type of Academic Research: Critique of Vulgar Raymondism', *First Monday*, 4.10 (1999), https://doi.org/10.5210/fm.v4i10.696

9 Alan Galey and Stan Ruecker, 'How a Prototype Argues', *Literary and Linguistic Computing*, 25.4 (2010), 405–24, https://doi.org/10.1093/llc/fqq021

share many properties with text and writing, indeed many more than most programmers and scholars usually acknowledge. When a programmer writes software, the result is not merely a digital object with a specific computational function. It is a program that can be executed by a computer, but, as so-called source code, it is also a text readable by humans (primarily, but not exclusively, programmers).[10] In the case of codework in humanities research, this text is also a part of an encompassing and larger epistemological framework comprising research design, theory, activities, interactions, and outputs. In the digital humanities context, the code part of this framework arises from a combination of the programmer's technical skills, her theoretical background knowledge (concerning both the humanities topic and computational modelling), and interpretations of the conversations she has had with collaborators, both academic and technical. It follows that, from an epistemic point of view, the practice of the programmer is no different from the practice of the scholar when it comes to writing.[11] Both are creating theories about existing epistemic objects (e.g. text and material artefacts, or data) by developing new epistemic objects (e.g. journal articles and critical editions, or code) to formulate and support these theories. In this sense, our view connects back to Bernard Cerquiglini's position that the scholarly editions of texts are not mere re-representations of some existing textual content, but *theories* about that content.[12]

The analogy we draw between code and programmers, on the one hand, and print publications and scholars, on the other, parallels Bruno Latour's comparison of machines and engineers, with texts and writers.[13] In relation to the practice of developing machines, and their application in scientific research, Latour also makes reference to the

---

10   Moritz Hiller, 'Signs o' the Times: The Software of Philology and a Philology of Software', *Digital Culture and Society*, 1.1 (2015), 152–63, https://doi.org/10.14361/dcs-2015-0110

11   Joris J. van Zundert, 'Author, Editor, Engineer: Code & the Rewriting of Authorship in Scholarly Editing', *Interdisciplinary Science Reviews*, 40.4 (2016), 349–75, https://doi.org/10.1080/03080188.2016.1165453

12   Bernard Cerquiglini, *In Praise of the Variant: A Critical History of Philology* (Baltimore, MD: The Johns Hopkins University Press, 1999).

13   Bruno Latour, 'Where Are the Missing Masses, Sociology of a Few Mundane Artefacts', in *Shaping Technology-Building Society. Studies in Sociotechnical Change*, ed. by Wiebe Bijker and John Law (Cambridge, MA: MIT Press, 1992), pp. 225–59, http://www.bruno-latour.fr/node/258

idea of the 'black box', which he defines as any technology, instrument, theory, or algorithm that is considered to be so well established as fact that it is beyond question; scientific controversies surrounding the construction of the 'black box' have arisen, been resolved, and become effectively invisible.[14] In Latour's explanation of science as a social act, the construction of 'black boxes' allows larger epistemological constructs to develop; by the same token, controversies in science can be understood as attempts to construct and defend, or attack and destroy, particular 'black boxes' in the making. A 'black box' thus comes into being precisely through the establishment of trust in its correct functioning, which is done by seeking a consensus about its correctness within the bounds, and according to the social mechanisms, of the scientific community.

In the humanities, however, the term 'black box' is often used to signal some unknown: a theory or instrument that has not undergone critical inspection and cannot, therefore, be trusted. Thus the labelling of a particular software technology as a 'black box' has come to mean, in some parts of the humanities, precisely the opposite of what was intended: rather than signalling that 'this is a trusted instrument', it signals 'this is an instrument which is suspect, and deserving of critical attention.'[15] Arguably the perverse implications of the label, and the

---

14   Bruno Latour, *Science in Action: How to Follow Scientists and Engineers Through Society* (Cambridge, MA: Harvard University Press, 1988).

15   See, for instance, Max Kemman, Martijn Kleppe, and Stef Scagliola, 'Just Google It: Digital Research Practices of Humanities Scholars', in *Proceedings of the Digital Humanities Congress 2012*, ed. by Clare Mills, Michael Pidd, and Esther Ward (Sheffield: HRI Online Publications, 2014), http://www.hrionline.ac.uk/openbook/chapter/dhc2012-kemman: 'Google introduces a black box into the digital research practices of scholars, but interestingly enough this does not seem to influence the trust of the majority of scholars in search results'; also, P. Svensson, *Big Digital Humanities: Imagining a Meeting Place for the Humanities and the Digital* (Ann Arbor, MI: University of Michigan Press, 2016), https://doi.org/10.1353/book.52252, http://hdl.handle.net/2027/spo.13607060.0001.001, talks about 'the importance of providing material results to the users rather than quantitative "black boxes" results' (p. 92). Svensson, interestingly, also uses the metaphor for the organisational mechanisms of the globally overarching organisation for digital humanities, ADHO (Svensson, *Big Digital Humanities*, p. 79). Johanna Drucker, although not specifically using the metaphor of 'black box', talks about 'reification of misinformation' when addressing computational quantitative measures on data we cannot see, with provenance we cannot verify, using algorithms we do not know (Johanna Drucker, 'Should Humanists Visualize Knowledge?', *Vimeo*, video lecture at Lehigh University, Bethlehem, Pennsylvania, 2016, https://vimeo.com/140307034).

suspicion with which so-called 'black boxes' are treated, are precisely the symptoms of the failure of the existing social, scholarly mechanisms, within those sectors of the humanities that are most distant from the empirical end of the science spectrum, to incorporate instruments and theories that arise from without. The result is a poignant mutual incomprehension: those who create software often understand their goal precisely to be the construction of a (trustworthy) 'black box', and they draw upon the mechanisms of science to do so — for what programmer wishes her code to be considered untrustworthy? And yet this very attempt to build the trust necessary for the instrument to attain 'black box' status, especially if the attempt is accompanied by the sort of discourse common in the empirical sciences, causes distrust in a community where consensus and dissent work differently.

Put another way: the very qualities and practices that, in other contexts, would create trust in software tools, now tend to diminish trust in them in the humanities context. In order to begin to counteract this paradox we can perhaps draw on the idea of code as an argument. As Richard Coyne and David Berry, among others, have shown, the internal structure and narrative of code ought not to be regarded as a mathematically infallible epistemological construct, although formal and mathematical logic is involved in its composition, just as logic has a natural place within rhetoric.[16] If we consider code as a rhetorical rather than a mathematical argument, it parallels humanities knowledge production in terms of theory and methodology. Code can thus inherit the multi-perspective, problematising nature and diverse styles of reasoning that are particular marks of methodology in the humanities. From this perspective, different code bases represent different theories, each of which needs to show its distinctive, true colours in order to be adequately recognised and evaluated.

Until now, however, most fields within the humanities lack a system for approaching, in a critical fashion, any argument that code presents, and for evaluating the workings of software. The discourse critiquing and evaluating code in the (digital) humanities has mostly focused on tenure track evaluation and peer review of the 'surface' of digital

---

16 Richard Coyne, *Designing Information Technology in the Postmodern Age: From Method to Metaphor*, A Leonardo Book (Cambridge, MA: MIT Press, 1995); Berry, *Critical Theory*.

objects, i.e. the resulting interface or visual presentation.[17] Within the humanities, very little work has been done on practical code review or on the evaluation of the inner logic of code.[18] To this end, some work has been done in new media and software studies, especially where software has a role as a production tool of cultural artefacts in film, art, and so forth.[19] This work, however, primarily concerns itself with the 'theoretical discussion of how software interacts with society, influencing our perception of the world'.[20] With some noted exceptions,[21] academic journal articles in the humanities rarely engage with the actual source code that underlies computationally-derived research results. A methodological examination essentially restricts itself to the results obtained from a graphical interface, or the interpretation of the quantitative results generated by a software program. A typical paper might report what statistical measure had been used, but generally omits to mention which software was used to make the measurement; in the case of project-specific software, the quality of its implementation is not examined. Many of the standard mechanisms for quality control in the software industry, such as line-by-line code review, unit testing, regression testing, and measurement of the extent to which the tests are comprehensive ('code coverage'), are routinely omitted in the humanities programming context, including in larger projects and even some centres.[22] Yet it is this type of engineering knowledge (crucial as it

---

17 Susan Schreibman, Laura Mandell, and Stephen Olsen, 'Introduction', *Profession* (2011), 123–201, https://doi.org/10.1632/prof.2011.2011.1.123; Kathleen Fitzpatrick, 'Peer Review, Judgment, and Reading', *Profession* (2011), 196–201, https://doi.org/prof.2011.2011.1.196

18 Joris J. van Zundert and Ronald Haentjens Dekker, 'Code, Scholarship, and Criticism: When Is Coding Scholarship and When Is It Not?', *Digital Scholarship in the Humanities* (2017), https://doi.org/10.1093/llc/fqx006

19 E.g., Manovich, *Software Takes Command*; and Mark C. Marino, 'Field Report for Critical Code Studies, 2014', *Computational Culture*, 4 (2014), http://computationalculture.net/article/field-report-for-critical-code-studies-2014%E2%80%A8

20 Chiara Bernardi, 'Working Towards a Definition of the Philosophy of Software', *Computational Culture*, 2 (2012), http://computationalculture.net/review/working-towards-a-definition-of-the-philosophy-of-software

21 For instance, *Cultural Analytics*, https://culturalanalytics.org, and *Computational Culture*, http://computationalculture.net/ could be mentioned. However, even in issues of these publication platforms, which are geared specifically towards critical engagement with data and software, one searches in vain for actual source code criticism.

22 Although we do not wish to call out specific examples of projects or tools that omit these practices, because the problem is so widespread, the reader is invited to

is in establishing the correct working of the code and gauging its inbuilt assumptions) that would be fundamental to the critical examination of the software that is applied in humanities research. Perhaps the fact that software and code peer review do not count towards academic credit[23] in most academic contexts plays into this state of affairs. There is no incentive for humanities researchers to consider the scientific or technical quality of the software tools they wield, nor is there sufficient training to acquire the skills to do so. Software engineering in the humanities ranges from professional teams working in conformance with industry testing best practice, to untested one-off scripts created by individuals. The scholars who rely on these tools lack the means to gauge the quality of either.

As code is an increasingly important epistemic object in humanities research, the state of affairs described above creates a real methodological problem; this gives rise to an urgent need for a practical examination and theoretical discussion of how software reflexively interacts with humanities research. We contend that both code as an epistemic object, and codework as an epistemic practice, must be given proper theoretical and methodological recognition in the digital humanities, along with the consequences and the rewards that such recognition bears. The current practice of 'black-boxing' the code results in a neglect of its epistemological contributions, and imperils one of the key components of knowledge production in the digital humanities.

There are three steps in particular that could be taken towards solving the deficiencies in current peer review practices concerning code and codework. First, there is a need for peer review and the critical examination of source code itself.[24] Second, open publishing of code in verifiable ways is already easily facilitated through existing public repositories such as GitHub and SourceForge, or institutionally-

---

peruse the code bases of those tools and projects that have been made open source, and to reflect on the fact that quite a bit of software in the humanities is not open source at all. The authors have frequently heard 'I would be embarrassed for others to see the code' cited as a reason for keeping source code in humanities projects closed.

23 Cf. again Schreibman, Mandell, and Olsen, 'Introduction'. For a particular poignant case consult Sean Takats, 'A Digital Humanities Tenure Case, Part 2: Letters and Committees', *The Quintessence of Ham* (7 February 2013), http://quintessenceofham.org/2013/02/07/a-digital-humanities-tenure-case-part-2-letters-and-committees/

24 Cf. also, again, van Zundert and Haentjens Dekker, 'Code, Scholarship, and Criticism'.

run versions thereof; but, in addition to this, its proper citation must become common practice in the humanities.[25] Third, reflexive accounts of (digital) humanities codework and ethnographic studies of actual work can help us understand how code and codework are changing the humanities.[26] The current contribution focuses primarily on this latter type of work by following and observing the experience of two digital humanities' programmers in order to derive insights and recommendations for those whose work may be affected by, or related to, codework in the humanities.

Methodology

The concept of the 'black box' can be seen as a methodological notion that is helpful in differentiating between the 'process' and the 'output' of knowledge production. In Latour's words, '[I]f you take two pictures, one of the 'black boxes' and the other of the open controversies, they are utterly different. They are as different as the two sides [...] of a two-faced Janus. "Science in the making" on the right side, "all made science" or "ready-made science" on the other.'[27]. 'Black-boxing' can thus be perceived of as a process of enclosing the tumultuous complexity of epistemological and methodological dilemmas, controversies, compromises, and decisions that are visible in the process yet hidden in the output of knowledge production.

Our objective in this chapter is to apply Latour's first rule of method to the socio-technical context of creating an argument through codework: we will examine scholarship in the making, and follow and reopen the dilemmas and controversies of the process of knowledge production before they get enclosed in the 'black box'. We thus reopen and analyse the process of DH codework, that is, we look at the inner practices, dilemmas, and decisions of programmers as they do their work. To do this, we have used the analytical autoethnography method, which Leon

---

25 Juriaan H. Spaaks, 'The Research Software Directory and How It Promotes Software Citation: Improve the Findability, Citability, and Reproducibility of Research Software', *EScience Center* (11 December 2018), https://blog.esciencecenter.nl/the-research-software-directory-and-how-it-promotes-software-citation-4bd2137a6b8

26 Christine Borgman, 'The Digital Future Is Now: A Call to Action for the Humanities', *Digital Humanities Quarterly*, 3.4 (2009), www.digitalhumanities.org/dhq/vol/3/4/000077/000077.html

27 Latour, *Science in Action*, p. 4.

Anderson defines as 'ethnographic work in which the researcher is 1) a full member in the research group or setting, 2) visible as such a member in the researcher's published texts, and 3) committed to an analytic research agenda focused on improving theoretical understandings of broader social phenomena'.[28]

In our case, the analytical autoethnography unfolded in the context of a research group that consisted of three members: two DH programmers (Andrews, and Van Zundert) and one ethnographer (Antonijević). The composition of the research team enabled us to engage in a study that combined autoethnography and collaborative ethnography.[29] In practice, our study had the following methodological design:

1. The team's ethnographer formulated a set of ten questions (see Appendix 6.A) aimed at generating reflexive accounts and examples of DH codework.

2. Each of the team's DH programmers individually answered questions on a written form, providing elaborate, semi-formal accounts of his or her DH programming practice.

3. The written accounts that were generated became the basis for a series of team discussions, both written and oral, which eventually formed the 'Experiences' section of this contribution.

This method enabled us to return from the final outputs of DH codework to the scholarly uncertainties and resolutions that preceded them. Through this reconstruction we were able to document some of the key phases in the epistemological construction of code artefacts, and to identify methodologically significant moments in the stabilisation of those artefacts. In other words, we relied on the experiences of two DH hybrids: scholars proficient in both humanities research and coding, who were seeking to make explicit what DH coders themselves know, perhaps tacitly, about why and how they code. In this chapter, we have removed titles and other similar identifiers of the projects and

---

[28] Leon Anderson, 'Analytic Autoethnography', *Journal of Contemporary Ethnography*, 35.4 (2006), 373–95, https://doi.org/10.1177/0891241605280449

[29] Cf. L. E. Lassiter, *The Chicago Guide to Collaborative Ethnography*, Chicago Guides to Writing, Editing, and Publishing (Chicago, London: University of Chicago Press, 2005), http://bit.ly/2iLCmGY

software that formed the basis of our autoethnographic accounts in order to protect the privacy of the colleagues and institutions related to these projects. The written autoethnographic accounts have been quoted in their original form, except for being slightly shortened and edited for clarity.

The goal of our methodological approach was twofold: 1) enabling programmers to develop a method through which they can reflect on their practices, understand them better, and communicate them to others, and, 2) providing traditionally trained humanists with a systematic insight into the inner epistemological and methodological workings of coding in the digital humanities. As mentioned previously, we do not aspire to be representative of DH codework, but to open a debate about some of the hidden elements in this practice. Combined, these two goals could offer a better understanding of codework as an activity of knowledge production in the humanities, along with criteria for evaluating, challenging, and/or rewarding those activities. Our approach thus addressed the challenge of making codework visible again in order to understand its ontology, origin, and effects.[30]

We have grouped our observations into the categories known as 'the five canons of rhetoric' (as proposed by Cicero in his *De Inventione*): *inventio, dispositio, elocutio, memoria,* and *actio*. Although originally developed for public speaking, these canons have proven to be an equally potent heuristic in analysing written and, more recently, digital discourse.[31] Our contribution seeks to extend this heuristic to the analysis of coding as argumentation, not in an attempt to fit codework and its elements into a pre-defined ontology, nor to suggest that it fully conforms or matches classical rhetoric. Rather, it is a way of presenting our experiences and claims in a form that we expect will facilitate interpretation by scholars who are well-versed in text production but likely less so in codework.

All three authors have worked in professional IT contexts as part of teams that worked according to formal software development

---

30   Cf. Berry, *Philosophy of Software*.
31   Laura Gurak and Smiljana Antonijević, 'Digital Rhetoric and Public Discourse', in *The Sage Handbook of Rhetorical Studies*, ed. by Andrea A. Lunsford, Rosa A. Eberly, and Kirt H. Wilson (London, Thousand Oaks: SAGE Publications, Inc., 2009), pp. 497–508, https://doi.org/10.4135/9781412982795.n26

methodology (including, for example, iterative development, unit testing, code reviews, continuous integration, automatic builds and deployment, etc.). The two authors who served as subjects for the study both have dual backgrounds as formally trained academic researchers and as professional programmers. Both authors have also created and wielded bespoke code as individual programmer-researchers and as members of teams in an academic context. Their 'hybrid' skills and experience therefore make them excellent candidates to compare various types of development and academic engagement with software and source code.

Experiences

Inventio — The Impetus for DH Researchers to Code

The *inventio* stage of scholarly programming is usually driven by a specific research need: to collect data, to see a set of data in a different way, to (try to) answer a research question; to develop a new method; or to tweak some of the existing tools, resources, and/or data so as to adjust them to one's specific research needs or workflow practices. There are also other catalysts, such as being hired to do DH programming on a project, doing one's own 'free floating stuff' (as will be discussed below) and playing with technology, mastering new tools and skills, and so on. This 'spark' of invention sets off a generative process — building, tinkering, tearing down and rebuilding — that goes on until the programmer understands the parameters of the challenge.

In many cases, a humanities-specific research question will drive the software development and coding. A research design is formulated in a dialogue between developer and researcher, and it demands a workflow that can be expressed or operationalised by a developer within digital media. A particular question might be, for instance, which parts of a particular text were written by different authors: 'I searched for applicable author identification methods (which were more related to statistics than to coding) [...] those methods were then "poured" into a code form for practical tests.'

Codework yields its own reflexive research questions as well, which may initiate new research and new code.

> I think [this project] is a good example. It is an attempt to find a way of coding that is closer to close reading and hermeneutics than big data analysis. [The] intent [is] to explore different modes of coding that are closer to humanities-style reasoning. For me [it is] the most intimate way of trying to find how coding is a humanities literacy.

Obviously, codework involves more than merely coding technology. Just as other types of researchers may resort to schemas, index cards, photographs, and thesauri, coding is not a single-instrument creative activity. Coders use interrogation, dialogue, drawings, and schemas in an attempt to come to a close understanding of the domain and concepts that researchers apply.

> There have been other situations where the purpose of my programming was to reverse-engineer and replicate the model of data I was given. Unfortunately the only clue I had as to the intended data model was the website that was built around the data, which means that I had to do a lot of trial-and-error guessing [...] A large part of this 'programming' task was to get a big sheet of butcher paper and make an enormous diagram by hand, recording the connections between the database tables as I figured them out, and unearthing thereby the queries that were hidden on the web server. [...] when I say 'hidden' I don't mean 'obscured in illegible source code' — I mean that I actually had no access to the code that contained them.

Coding can also be a means of learning and testing new skills and methods. A distinction could be made between skill-gathering projects and research projects. In research projects, the development of code will be driven by a research question, and the developer will apply well known and rehearsed tools and techniques to the problem insofar as possible. Conversely, skill gathering is driven by the need to explore and examine new tools and techniques. Such projects need not lead to actual research results, 'but coding in this sense is a good way of keeping your code skills up to date. If you're lucky enough you might draw a small paper out of that kind of coding that really is training.' In this sense coding-to-learn is equivalent to scholars keeping up with, for instance, publications in critical theory or factoring a new-found approach into an argument about the sources they are working with.

Often, method development and new research insights will co-evolve during code development:

Perhaps the most 'scientific' programming I have done is the work on the [xyz] software. That project has been much more about trying new methods than about improving established ones, and so had a different character from the outset. I had to decide how the data ought to be collected and represented; I had to constantly revisit these decisions as I collected data that challenged my previous assumptions and heuristics. I had to discuss some concepts with computer scientists who understood more about graph arithmetic than I did in order to explain what I was trying to accomplish. The 'meat' of [this project] is a sort of calculator that, given a stemma[32] and given a set of textual variants, colors each manuscript within the stemma according to which textual variant it contains, and then works out whether that particular pattern of colors (that is to say, text mutations) could possibly have descended in a genealogical way. In a sense this is not new at all — every textual scholar understands what 'genealogical' variation implies, in the sense that they have expectations about which manuscripts in a stemma ought to share that change — but in another sense it is entirely new, since common scholarly wisdom held that there is not a lot you can do with a 'contaminated' stemma (that is, a stemma that indicates that a manuscript was copied by comparing or mixing several exemplars), but the computational model that my CS [computer science] collaborators and I developed can treat 'contaminated' stemmas in exactly the same way as traditional ones.

Thus, the argument that code begins to construct is grounded in well-established textual theory and methods. In this case, the argument is based on (parts of) the stemmatic approach (often also known as Lachmann's method), which is used to establish the genealogy of manuscripts based on variant readings that are accrued when manuscripts are copied over time. Similar to the more traditional scholarly article, the code expresses and uses these existing humanities methods and builds new methods and argument from there.

Not all code starts out with high research aspirations. Essentially, code is always used to automate work that would otherwise have been done manually, or would have been too onerous to do at all. These can be very simple tasks, such as writing a script to highlight changes between drafts of a paper, or very complex things such as writing a generic text collation tool. The development of the software is driven by tasks specific to the research at hand. But the need for coding is also born from the computational workflow itself and the need to move

---

32 A stemma is a tree-like representation of the genealogy of documents that represents an assertion about how later documents were copied or derived from earlier ones.

data from one data model to another. Thus, software use leads to more software needs: 'I use [this particular] web software for transcription of manuscripts, but in order to do anything with the data after I've transcribed it I need to be able to extract it from [this web software] in the form I need.'

Even though code may be geared towards facilitating tedious and repetitive, simple tasks within a research workflow, interesting and complex research designs will likely be more stimulating to developers than mundane support tasks purged of their direct relevance to the research question, for example, data preprocessing: 'Nowadays, being a senior researcher [...] I will code [...] when I can work from a clear research question and not from some derived coding directive.'

Related to the use of programming as a means for acquiring new methodological skills is the idea of building code as play and tinkering, an idea congruent with Geoffrey Rockwell's characterisation of text analysis and research as a form of disciplined play.[33]

> [Doing] free floating stuff. That stuff wasn't driven by research questions though. It was more solutions looking for a problem. There were these interesting text analysis methodologies and techniques, impressive statistical approaches to stylometry, etc., that just made my fingers tingle to get hands-on and to apply them to concrete problems. A friend of mine called this 'haptic thinking', a way of developing thoughts and new insights through using your keyboard.

This tinkering and play may sometimes be criticised as 'not research-driven enough', but it can actually yield very interesting results, and points to new ways of looking at a problem. However, in certain contexts, this does get recognition: for instance, some digital humanities centres give programmers a day off to pursue their own ends.[34]

Dispositio — How Coding Constructs Argument

Like text authorship, codework often consists of writing and re-writing, as well as the configuration and reconfiguration of larger pieces of code.

---

33 Geoffrey Rockwell, 'What Is Text Analysis, Really?', *Literary and Linguistic Computing*, 18.2 (2003), 209–19, https://doi.org/10.1093/llc/18.2.209

34 Smiljana Antonijević, *Amongst Digital Humanists: An Ethnographic Study of Digital Knowledge Production* (London, New York: Palgrave Macmillan, 2015), https://doi.org/10.1057/9781137484185

A programmer works by writing lines of code and combining these into larger, meaningful constructs — not unlike how lines and paragraphs of prose come into being. As with writing, much of coding's activity is to restructure individual lines of code and functions[35] until a satisfactory behaviour is achieved. This restructuring takes place at many levels: lines of code, functions, groups of such functions (called modules or libraries), and entire applications and their constituents — for example, databases, web frameworks, program core, file systems, and security layers. Many decisions are made on how to arrange these pieces while the codework is ongoing. These decisions are informed by experience, knowledge of the research domain, and considerations of feasibility, performance, and resources. They often rely as much on assumptions or educated guesses as on concrete knowledge.

Technical decisions are a necessary part of any code development trajectory. Yet, programmers learn from experience that their decisions will often change. 'A number of decisions need to be made in advance — what kind of database will I use? Is there a programming language that is particularly suitable to the task at hand, or can that decision be arbitrary? However, these decisions are essentially never final.' Technologies are swapped in and out for many reasons: performance, technical innovation, convenience of programming: '[This project] began life backed by a relational database, and then was moved to an object datastore, and is now on its way to migration into a graph database. The software was written in Perl, but its graph-database replacement is being written in Java.' Such technical decisions can affect the methodological make up of research, and it takes expertise in both coding and research design to judge them.

Thus, there is more to these decisions than purely technical considerations: a great deal depends on assumptions about and factuality of input data and research design.

> One thing that comes with experience as a programmer is the understanding that, to the extent that you do not wield perfect control over the information that is the input to your code, you are (or another developer is) probably at some point going to have to change how your code models and processes that information. This applies as much in theoretical physics or commercial software engineering as it

---

35 A set of the lines of code that fulfil a discrete function.

does in the humanities; it's just that one often encounters this need for adaptation very rapidly within the humanities [...] The sorts of simplistic 'shortcuts' that are common in industry or in computing in the natural sciences tend not to have a lot of useful longevity in code bases in the humanities.

This last statement in particular reveals one way in which coding in humanistic contexts tends to be distinct from coding in other domains. Humanities research deals with strongly heterogeneous data; given historical and cultural context, the importance of the situatedness of information has strong ramifications for the models and processes that are applied by the code.[36] Where the sciences may abstract away from particulars to allow patterns to emerge, the particular (the exception) is often precisely what the humanities scholar seeks. As, for instance, one of our programmers recounted: 'when you are building a prosopographical database you are not starting from formal definitions of what a "person" is and what its properties are, because "everybody knows" what a person is.' Objects and categories in the humanities are usually not as rigorously defined in their properties and attributes as are objects in the natural sciences, such as atoms or electrons. Who is an immigrant and who is native, for instance, largely depends on time, context, perspective, and who does the defining. Text is not a single stream of characters, but a complex object of layered signs and meanings, gender is far from binary, and borders of countries shift through time and geography.

Although decisions are perhaps never final, the decisions that are made can have far-reaching implications. These ramifications may occur at the level of the analytical design. For example, will a relational database be used or will a document store be applied? This choice corresponds to a primarily metadata-focused or object-focused approach. But decisions may also have institutional effects: should the software be unique bespoke code to be used only once by a single researcher, or is there an audience to be considered; and will continuous online availability have to be ensured? Such choices also lead directly to

---

36   Jackson, Virginia, and Lisa Gitelman, 'Introduction', in *'Raw Data' Is an Oxymoron*, ed. by Lisa Gitelman, Geoffrey C. Bowker, and Paul N. Edwards (Cambridge, MA: MIT Press, 2013), pp. 1–14, https://doi.org/10.7551/mitpress/9302.003.0002; Johanna Drucker, 'Humanities Approaches to Graphical Display', *Digital Humanities Quarterly*, 5.1 (2011), http://digitalhumanities.org/dhq/vol/5/1/000091/000091.html

decisions about life cycle management, maintenance, user support, and all the resources and management these demand.

What code will be written and how it is constructed also greatly depends on estimations of feasibility.

> If [a research design] is technically infeasible, if it is something which can't be meaningfully computed, there's no use trying. Similarly if the data is just not there or unattainable. But even if those prerequisites are met, then there's the question if it is feasible to code a solution in the time and with the resources available.

Decisions surrounding estimates of feasibility, research design, and code implementation are all comparatively informed: 'You will conjure up some of the latest on logistic regression and see if there have been similar questions, solved in similar ways, and this gives you good clues as to what and how you might do, build, and analyze.' Re-use, recombination, and reconfiguration lead to new methods and new code:

> Mostly we recycle existing ideas and we add a tiny new edge, application, or relevance to them. It is for this reason that I get suspicious if I really can't find a similar example of what I'm looking for, because 'new' mostly means a combination of what already went before but wasn't applied in a different context.

Here we see again that codework in the digital humanities follows epistemological principles that are equivalent to those in other forms of knowledge production, relying on continuous intellectual exchange with the community of practice. This is also observable in the re-writing of code. As in scholarship, argument by code is evaluated, changed, and re-evaluated in order to let it evolve into an acceptable scientific contribution. Confronted with real world data and real world use, programmers will quickly notice that many of their initial assumptions about the data, the model, and the process do not align with reality: 'Thus one gets into an iterative mode of rewriting, reworking or refactoring the code until it represents what it should represent and does what it should do.'

Both the programmers in our study feel that the most domain-relevant choices, i.e. the choices that are most pertinent to the research design and content analysis, are made in the so-called model. In codework, two types of models are usually differentiated: the data model, and the conceptual or domain model. The first deals with the technical aspects and should ensure safe storage, interoperability, performance, and so forth. The latter model pertains to the contents, the data as meaningful concepts, and the analytical part of the research. Ideally, this model applies an idiom that mimics the concepts that are native to a (research) domain: 'it is not unrealistic to say that this conceptual model is a simulation of the research process, or the analytics in real life. In my case definitely the more relevant decisions are made in this phase. Defining, tinkering with, and exploration-wise building that model.'[37]

Even if programmers and researchers alike tend to feel that attention to the conceptual model is the most relevant part, it is generally not where most of the effort demands to be directed. As in so many fields, a tremendous amount of time is spent in data gathering and curation: 'I think a very good deal — it's like an 80/20 rule — of coding effort goes towards handling and transforming data, and usually only a lesser bit of code and coding is spent on actual analysis.'[38]

Elocutio — Coding Style, Aesthetics of Code

Software code is a thing written, and, as much as the formal constraints of computer language allow, a software author has her own style, both in regard to the aesthetics of the code and the way of working to create the code. Every coder has a personal experience of *technē* that is essential to her methods. The author may subscribe to certain methodologies, such as Agile Software Development, and she may have particular aesthetic values in coding that may be connected to such mundane things as the use of tabs instead of spaces, but these values may also relate to how program control flow and conceptual composition is used to express research domain concepts and analysis in code. Personal aesthetics can also pertain to the choices made between functionally equivalent pre-existing libraries of code and applications that are reused by the developer.

---

37 On modelling and its complicated relation to digital humanities work and coding, see also Willard McCarty, *Humanities Computing* (Basingstoke: Palgrave Macmillan, 2005); and Julia Flanders and Fotis Jannidis, *Knowledge Organization and Data Modeling in the Humanities* (2015), http://www.wwp.northeastern.edu/outreach/conference/kodm2012/flanders_jannidis_datamodeling.pdf

38 M. Arthur Munson, 'A Study on the Importance of and Time Spent on Different Modeling Steps', *SIGKDD Explorations*, 13.2 (2011), 65–71, https://doi.org/10.1145/2207243.2207253

A scientific discipline is often characterised by a certain dominant style of research, writing, and publication.[39] In the humanities, for instance, the monograph is generally viewed as the most valuable form of publication,[40] and an individualistic intellectual approach is preferred.[41] In a similar way there are aspects of preferred style and form to the writing of code, both individualistic and as a norm in coding communities.[42] Code writing is not a clinical process of assembling discrete mathematical logic statements; aesthetics of the code itself and a feeling of 'being in the flow' exist in code authoring just as they exist in text authoring: 'Part of the story is a feel that's somewhere in between art and craft that accompanies coding. The very feel of building, of the keyboard rattling, of code lines getting formed on the screen and those doing something. Of working your way to a working algorithm or working tool.' Yet this particular feel of flow and art in the act of building is hard to describe: 'It feels like describing the color purple.' This feel is part of the personal style of working and the personal 'poetics' of code, which is important to adhere to. Neither of the study's programmers, for example, like so-called pair programming, even though evidence exists[43] that collaborative working on code is more effective and leads to less distractions or errors, and to a boost in efficiency: 'Pair programming doesn't work for me, just like dictating doesn't work. I can't simultaneously think up complex conceptual thoughts and turn them directly and unerringly into well-formed sentences.'

Work on the interfaces for software tools intersects with the programmer's style of work and style of coding. Although neither of the programmers are very enthusiastic about interface work (see also

---

39 Alistair Cameron Crombie, *Styles of Scientific Thinking in the European Tradition: The History of Argument and Explanation Especially in the Mathematical and Biomedical Sciences and Arts* (London: Duckworth, 1995).

40 Peter Williams et al., 'The Role and Future of the Monograph in Arts and Humanities Research', *Aslib Proceedings*, 61.1 (2009), 67–82, https://doi.org/10.1108/00012530910932294

41 Wolfgang Kaltenbrunner, *Reflexive Inertia: Reinventing Scholarship Through Digital Practices* (Leiden: Leiden University, 2015).

42 E.g., Guido van Rossum, Barry Warsaw, and Nick Coghlan, 'PEP 8 — Style Guide for Python Code', *Python* (5 July 2001), https://www.python.org/dev/peps/pep-0008/

43 Charlie McDowell et al., 'The Impact of Pair Programming on Student Performance, Perception and Persistence', in *Proceedings of the 25th International Conference on Software Engineering*, ICSE '03 (Washington, DC: IEEE Computer Society, 2003), pp. 602–07, http://dl.acm.org/citation.cfm?id=776816.776899

the section '*Actio*') as it diverges all too soon from a research focus, it does affect the way in which the programmer uses code to develop argument. One programmer stated:

> I do like interface work as long as it is aimed at this exploring new modes of being for text, but as soon as I have to start to take care of a real user base the questions diverge from my actual research question pretty soon.

While the other programmer put it as follows:

> It does affect the way I write code, not only because I have to think a little (or a lot) harder about interface and usability when I expect others to use it, but also because I have to spend a little bit of time second-guessing how their assumptions and use cases might differ from my own. I try not to go too far in that, though — I find that engagement with real users and their needs when they actually appear is a more effective way to extend a tool than conjuring up hypothetical users and their needs.

Memoria — The Interaction between Code and Theory

We associate the rhetorical canon of *memoria* with the ability of code and codework to serve as memory systems that embed theoretical concepts within objects and recall them when needed, in order to augment research methodology and create new theory. In this way, the ability of code and codework to serve as memory systems parallels that of a book or a library. In the humanities, theory in both digital and 'conventional' fields has major and direct bearings on the programming and codework arising from these fields. We should seek, therefore, to illuminate and explain how exactly code embeds humanities theory and operates under its influence: this should not be hidden lest it be prematurely labelled a 'black box'. Similar to writing, code and coding are also an interpretation and reinterpretation of theory; like any narrative or theory, code is not some neutral re-representation: its author selects, shifts focus, expresses, and emphasises.

As methodologies go digital and their practitioners speak ever more in terms of 'data', critical theory remains fundamental to fostering understanding that there is no such thing as raw data, and that all data, including digital, is constructed, created, and situated.[44]

---

44 Jackson and Gitelman, 'Introduction'; Johanna Drucker, 'Humanities Approaches to Graphical Display'.

> Critical theory has a major bearing on aspects of the creation of a data model; if, for example, I were collecting a dataset in which I needed to record characteristics like 'race' or 'gender', I would have to think long and hard about how that information ought to be structured. I have run into this in the data of others, [for instance a] prosopography dataset and how it deals with the category of 'eunuchs'.

Theories from the humanities directly impact on the choice of tools and technologies that programmers use. Codework is far from theoretically uninformed, but rather theory driven.[45] One of our programmers explains that hermeneutic inference is not simply supplanted by scientific models. Rather, models evolve to express the complexity of the research object and to reflect theoretical-interpretative aspects: 'In one of our projects we started out with a very simplistic neural network that just took the vocabulary of novels as input. But to be able to correlate to readers' judgement of literary style we were soon integrating word2vec and doc2vec models to reflect theoretical notions like themes and perspective.' The other programmer explains, to illustrate, that a particular graph model used for text fits more naturally with certain arguments of post-structuralism than other models do. Transcribing a text through a model that does not assume a single, or even a single 'main', sequence of characters, is a coded reflection of an epistemological understanding that text is a multi-layered, multi-dimensional object of information, rather than a one-dimensional array of signals. When using a graph model, this programmer acknowledges making a certain set of claims about the text, and that these can be seen to be in line with post-structuralist arguments, or even with the tenets of new philology. However, this programmer also adds that such use of the model has more to do with its fitness for the research approach being tried, and less with a personal belief or conviction of what text 'should' be like: 'I am aware of that, as I use it, but at the same time I tend to want to avoid ascribing more significance to a particular computational model than it perhaps warrants.' It is important therefore to be aware of the risk of misreading the declarative nature of code: 'I think it is quite a common experience for DH programmers to have others ascribe much more

---

45 Jean Bauer, 'Who You Calling Untheoretical?', *Journal of Digital Humanities*, 1.1 (2011), http://journalofdigitalhumanities.org/1-1/who-you-calling-untheoretical-by-jean-bauer/

argumentative intentionality, presumption of declaration of authority, or sheer "staying power", to their models than they actually intended.' It is all too easy for programmers to be caught in this way between rival schools of thought in the humanities, and their code pointed to as evidence of a positive disregard for one critical theory or another.[46]

DH theory also adopts a reflexive stance. We may imagine, for example, a project that enters into a dialogue with the embedded ideology and discourse of big data approaches and markup languages. The discourse of the first is bound up in empiricism, quantification, scale, and speed; and thereby glosses over the precision of reasoning, the heterogeneity of data, the situatedness of data and data production, an abductive style of reasoning, and so forth, all of which are distinctive traits of many methodologies within the humanities. The discourse of big data and of markup languages is tied to certain epistemological theories: to quantification and scientism in the case of big data,[47] and to hierarchical epistemological structures and representational philosophy in the case of markup languages.[48] One of our DH programmers is engaged in a project that specifically aims to experiment with other, more hermeneutic styles of coding: 'Obviously in this case also these theories have a very direct influence in the code and coding style. I cannot, for instance, use something like machine learning unless I can convincingly argue that it serves some slow programming or close reading aspect. This is where I also still struggle very much.'

Coders take in theory and implement it. Code can be regarded as a performative application or the explanation of theory rather than a written account of it.[49] As one programmer put it:

> I guess the difference is that with code I can make it *do* something, thus I tend to try to argue through transformations of data. [...] So in both print and code I somehow argue about [the object of study]. In print I do mostly by abductive logic, that is, plausible reasoning, my reasoning or my evidence is a narrative. In code I reason by transformation, I think.

---

46 E.g., Jones, *Emergence of the Digital Humanities*, pp. 31–32.
47 Cf. Johanna Drucker, 'Graphesis: Visual Knowledge Production and Representation', *Poetess Archive Journal*, 2.1 (2010), https://journals.tdl.org/paj/index.php/paj/article/download/4/50
48 Steven J. DeRose et al., 'What Is Text, Really?', *Journal of Computing in Higher Education*, 1.2 (1990), 3–26, https://doi.org/10.1145/264842.264847
49 Cf. Adrian Mackenzie, 'The Performativity of Code', *Theory, Culture & Society*, 22.1 (2005), 71–92, https://doi.org/10.1177/0263276405048436

An added attractiveness of code is its meticulous performance; under the same conditions code will always operate the same way, yielding the same results. This is 'an affordance that allows [one] to build [an] argument in a very iterative and controlled way'. Each statement added to a body of code is a building block of a transformative system or workflow; each bit of transformation is a bit of 'evidentiary' or 'argumentative' performance.

Lastly, codework adds to theory, as we have witnessed in an example already given above:

> In a sense this is not new at all — every textual scholar understands what 'genealogical' variation implies [...] but in another sense it is entirely new, since common scholarly wisdom held that there is not a lot you can do with a 'contaminated' stemma [...] but the computational model that my CS collaborators and I developed can treat 'contaminated' stemmas in exactly the same way as traditional ones.

Although it is clear that code is in some form an argument and encompasses or expresses theory, we must also acknowledge that the argumentative rhetoric of code is very limited:

> Anything that you might call an 'argument' in my code is going to be pretty oblique, or going to be a passive argument by virtue of its inclusion in my model [...] academic writing tends to be expressed in rhetorical forms that are intelligible to readers, that advertise what I consider to be a fact beyond dispute, what I acknowledge as an unresolved argument but am nevertheless lending my support to one side or the other, and what I consider to be the original contribution of the argument I'm making. These rhetorical forms are entirely non-existent in code, and can only be replicated to some extent in comments and in documentation.

Actio — The Presentation and Reception of DH Codework

In codework, *actio*, the delivery of one's argument, could be compared to the publication and reception of the software and its source code. To date, DH programming is generally not recognised as a locus of humanities expertise. Coders who are trying to accumulate academic recognition and credit for their computing activities must make use of methods geared entirely toward the delivery of prose, such as publications and presentations. For scholars on the research track, most programming is done for themselves, for their own research needs, and thus with the self as the 'intended user'. This undermines, in a meaningful and constructive way, the assumption that DH projects should always be collaborative and that their programmers should work with other humanities scholars in mind, which foregrounds the view that programmers are 'technical staff' working on behalf of researchers. Programming is often seen as a technical activity and not as research, and DH programmers are consequently seen as 'technical problem solvers' whose competencies are outside the realm of humanities expertise. This assumption is particularly entrenched for junior hybrid scholars in the 'alt-ac' (alternative-academic) careers whose professional identities and paths are often ambiguous; scholars on traditional academic tracks (especially those appointed to teach digital humanities) are accorded more recognition insofar as they have built a publication record and engaged in other forms of academic visibility and acknowledgment (conferences, projects, etc.).[50] This lends credibility to hybrid scholars, even if their traditionally trained colleagues do not fully grasp their work.

As with any other work, codework is shaped and influenced by its (social) context, which may positively or negatively influence the attitude and perception that coders hold towards their work. Both of our programmer-scholars have experienced such positive and negative influences in industry as well as in academia. One of them recalls a specific instance of a severe disconnect between management, researchers, and computer engineers. None of the groups understood much of the others' methods, motivations, commitment, or particular needs as to incentives and rewards, and were therefore unable to work very productively towards the shared research aims. This resulted in 'a lot of frustration', and a dislike, in the case of the programmer, for 'large and overcrowded' research projects. The salient point made by both programmers is that healthy interaction with others (be they co-programmers, other researchers, or management) is essential for inspired and productive research projects. Codework is very much interdisciplinary work that thrives on interaction. Both our programmers have worked in projects that had balanced and unbalanced research

---

50 Cf. Bethany Nowviskie, 'Where Credit Is Due: Preconditions for the Evaluation of Collaborative Digital Scholarship', *Profession* (2011), 169–81, https://doi.org/10.1632/prof.2011.2011.1.169; *#Alt-Academy 01: Alternative Academic Careers for Humanities Scholars*, ed. by Bethany Nowviskie (New York: MediaCommons Press, 2014), https://libraopen.lib.virginia.edu/public_view/6395w715k

governance; and found projects in which responsibility, accountability, and credit for design and methodology were all shared equally between humanities scholars and technologists to be far more rewarding and productive than research where all constraints were put forward by one primary investigator (PI). These observations tie in with the work of Helen Burgess and Jeanne Hamming, who argue that codework may be perceived by scholars as less of an intellectual labour than scholarly reasoning and writing.[51]

It is still hard for those doing codework in the humanities to receive acknowledgement for the academic quality or character of their software. Neither programmer reported any hint of the possibility of their code being academically evaluated or peer reviewed as digital output. Instead, academic acknowledgement and credit must be gathered through conventional venues like journals, papers, and print publication.[52] Even these garner precious little credit for the software itself, since the value of codework is often overlooked: '[T]he PIs [...] almost never acknowledged in articles and presentations who did much of the work.'[53]

Some research-track programmers have managed to build a research record despite not often being acknowledged as a researcher:

> Through the years I have heard many variants of the implicit 'what you do is not research'. Someone exclaims 'But you are in IT' with a subtext of 'you're not researching'; another colleague says 'But what you do is [IT] infrastructure'; at a conference I am complimented for still being in academia as a programmer: 'You must be doing something right.'

Others have simply deployed their coding skills in the service of their own projects: 'I have been on a more classical academic tenure track. So I was never "someone else's programmer", well, not in academia anyway.'

---

51 Helen J. Burgess and Jeanne Hamming, 'New Media in Academy: Labor and the Production of Knowledge in Scholarly Multimedia', *Digital Humanities Quarterly*, 5.3 (2011), http://digitalhumanities.org/dhq/vol/5/3/000102/000102.html

52 Cf. Ryan Shaw, 'On Tenure and Why Code Can't Speak for Itself', *Ryan Shaw*, https://aeshin.org/thoughts/on-tenure/ (accessed 6 November 2017, unavailable at time of publishing).

53 Cf., for instance, James Smith, 'Coding and Digital Humanities', *James Gottlieb: Seeing What Happens When You Collide the Humanities with the Digital* (8 March 2012), http://www.jamesgottlieb.com/old/2012/03/coding-and-digital-humanities/

This inability to derive any acknowledgement for codework will eventually reflect on the work itself: 'It was very hard to derive some sense of pride and accomplishment from such projects. [...] there was a consistent dissatisfaction in such projects, even though the coding, the content bit was fun.' It will eventually drive DH programmers to work primarily on projects where they are the primary investigator or have an equally visible research position: 'I will code if the coding leads to an opportunity to publish, to learn new analytic skills and methods, and when I can work from a clear research question and not from some derived coding-objective.'

When codework is not seen as genuine research contribution, it also becomes difficult for programmers to truly involve themselves in research-level discourse:

> In the earlier instances when people regarded me probably mostly as an apt programmer I only joined the research team in a phase after the general research question and design had already been discussed and set. In those cases it was not the research question that was posed to me, but [...] a vague idea of 'a tool' for some purpose [...] I think people generally saw me as some technical problem-solver, an implementation person, digital technician [...], certainly not a researcher. [Later] researchers tended to draw me into projects earlier and started reflecting, bouncing thoughts about research questions with me.

DH programmers on a research track have little incentive to accept research support roles doing technical work on behalf of others' projects. Software development as a service does not count towards tenure and arguably contributes little to one's own research.[54] Their programming work is often bespoke code, written to meet their own needs and often not looking beyond that. This is clear when our programmers were asked about their audience. One reacted, 'It's really me. I implement the code that operationalizes my research question, or the computation-analytical part thereof', while the other stated: 'At the outset I am almost always programming for myself.'

---

54 Cf., again, Takats, 'Digital Humanities Tenure Case'.

Conclusions

Looking at codework from the perspective of these rhetorical canons enables us to ground this commonly overlooked research activity within the humanities framework, and to explore coding as argumentation. Our exploration showed that codework reflects humanistic discovery in that humanities-specific research questions drive coding, and tasks specific to the humanities research motivate software development. Similarly, crafting and organising code resonates with the development and arrangement of a scholarly argument, that is, a programmer writes lines of code and makes many decisions on how to arrange these pieces into larger, meaningful constructs that influence the epistemological and methodological structure of research. Our study also illustrated that, like any humanities scholar, an author of software has her own style in respect to the aesthetics of the code and in the way of working to create the code; and this style develops through both individual norms and the norms of coding communities. We also showed that, in parallel to books and libraries, code and codework serve as memory systems that embed theoretical concepts in order to augment research methodology and create new theory where code can be regarded as a performative application or explanation of theory. Finally, our ethnography has illustrated how codework *actio* compares to the publication and reception of software where DH programming is still not recognised as a locus of humanities expertise and it is hard for humanities programmers to have their code academically evaluated as digital output.

These findings illustrate that, while code and codework increasingly shape research in all fields of the humanities, they are rarely part of disciplinary discussions, remaining invisible and unknown to most scholars. This invisibility has several important consequences. First, we believe that the integration of digital scholarship into the 'humanities proper' will be at a standstill as long as methods of digital knowledge production remain mysterious, misconceived, and/or mistrusted 'carriers of alien epistemological viruses'.[55] Second, software tools become 'black boxes' in the pejorative sense, as decisions about their constitution are made, essentially, without discourse or oversight.

---

55 Cf. Alan Liu, 'Digital Humanities and Academic Change', *English Language Notes*, 47.1 (2009), 17–35, https://doi.org/10.1215/00138282-47.1.17

Third, 'the medium becomes the message' in the worst way: the entire being of the software and the craft that went into its making is reduced, essentially, to the user interface that is presented upon its execution — precisely the aspect that analysis oriented programmers are likely to be least concerned about — which introduces a form of 'screen essentialism'.[56] Furthermore, we have seen in the chapters above that code constructs argument, but arguably this is not clear at all for humanities scholars who do not have any coding literacy. Lastly, 'hybrid' scholars who function as DH programmers find the path to scholarly credit and recognition for their work unnecessarily difficult, and, moreover, lack avenues to systematically improve their codework through discussions with peers, exchange of best practices, or similar established methods of scholarly learning and collaboration.

Code, like text, is not self-explanatory. Software, as Joris J. van Zundert points out, contains two messages: one is received when the code is read, and the other when the code is executed.[57] Critical interrogation of the software must perforce analyse both sets of messages. Indeed, it can be too easy for scholars (whether they code or not) to verge too far toward a screen essentialism that equates the interface with the meaning of the code, or to go to the other extreme of code essentialism by asserting that code is merely another form of text to be read.

In this study we have categorised the experiences of two coders in the humanities under the familiar headings of classical rhetoric canons. This is not just a gimmick. Although Mark Marino 'would like to propose that we no longer speak of the code as a text in metaphorical terms, but that we begin to analyze and explicate code as a text, as a sign system with its own rhetoric', it remains very unclear what that rhetoric is, how it works, and what its functional elements are.[58] Following Donald Knuth, it can be argued that code is, for all practical purposes, another form of literacy, and one that is not all that alien from the literacy of text.[59] To understand its specific rhetorical character, though, we need more insights into actual coding practices. We hope we have shown that auto-

---

56   Matthew G. Kirschenbaum, *Mechanisms: New Media and the Forensic Imagination* (Cambridge, MA: MIT Press, 2008).
57   Van Zundert, 'Author, Editor, Engineer'.
58   Marino, 'Field Report'.
59   Donald E. Knuth, 'Literate Programming', *The Computer Journal*, 27.2 (1984), 97–111, https://doi.org/10.1093/comjnl/27.2.97

ethnographies, such as the above, play a viable role in creating such insights. Along with interviews, contextual inquiries, diaries, and other ethnographic research techniques, reflexive accounts can create a record of what actually occurs when humanists write code, thereby, yielding the information needed to understand the particular poetics, praxis, and rhetoric of code creation. We are specifically not recommending that 'everyone should learn to code'. Code nevertheless plays an ever greater, almost ubiquitous, role in culture and society; and a humanities without the capacity to critique it would be ill-prepared to investigate its own society and culture.

Recommendations

Humanities scholars will be reluctant to bring into their research anything that they feel is both beyond their capacity to understand, and insufficiently endorsed by scholars whom they trust. Therefore, a strategy is needed for making code and codework visible, understandable, trustworthy, and reputable within humanities scholarship. It must be comprehensive, both in the sense of accounting for the source code and the executed result of software, and by including all relevant stakeholders. Based on our autoethnographic observations, we provide here a set of recommendations for the various groups of stakeholders, from individual scholars to academic institutions and professional organisations. Our recommendations are in line with Stephen Ramsay and Geoffrey Rockwell's argument that the evaluation guidelines for assessing digital work usually fail to tackle the central anxiety related to DH programming, which is how to recognise and rate this work as humanistic enquiry and scholarship.[60] Where we diverge from their argument is in the focus on materialist epistemology, taken in the sense that digital artefacts should be able to communicate their underlying theoretical assumptions or claims on their own. Ramsey and Rockwell contend that such theoretical assumptions can be inferred either by using a digital artefact or through accompanying stand-in documentation. In their view, such documentation should be avoided as it reinforces the linguistic dependence of scholarly communication, diminishing the ability to communicate scholarship through artefacts. They further

---

60   Ramsay and Rockwell, 'Developing Things'.

argue that, although source code could be seen as 'provid[ing] an entry point to the theoretical assumptions of black boxes [...] it is not at all clear that all assumptions are necessarily revealed once an application is decompiled, and few people read the code of others anyway. We are back to depending on discourse.'[61]

While we agree that providing source code is not sufficient for understanding the underlying theoretical assumptions, we disagree on viewing the 'dependence on discourse' as a feature that relativises the epistemic and communicative capacities of code and codework. In contrast, we argue that the interdependence of code and text should be embraced as a means of acknowledging their distinctive yet corresponding methods of knowledge production and communication. Just as code enhances text, making it amenable to the methodological and epistemological approaches of digital humanities; so too does text enhance code, making it more visible and intelligible for the humanities community. We believe that theoretical discussions on codework should become an established trajectory in the humanities, along with the development of methods for documenting, analysing, and evaluating code and codework. It is through the advancement of such methods and the acceptance of codework as a valid topic of humanities deliberations that the central anxiety that is related to DH programming will be cast aside. Van Zundert, following Friedrich Kittler,[62] argues that code and text literacy are on the same continuum of literacies, and are epistemologies with slightly different semiotics, which makes them well suited to reflect on each other.[63] The problems arise when either one is subordinated to the other, or when codework remains epistemologically and methodologically unexamined.

Evaluating code and DH programming in a disengaged way would be similar to the literary criticism enacted on a novel without reading it, which, in literary criticism, would be absurd. Yet it is currently the practice to 'criticise' software and code based only on the journal article that was derived from it. As much as possible, coders should support the involved evaluation of code as opposed to its disengaged criticism. On the other hand, coders regard graphical interface work as having low

---

61 Ibid., p. 81.
62 Van Zundert, 'Author, Editor, Engineer'; Friedrich Kittler, 'Es gibt keine Software', in *Draculas Vermächtnis* (Leipzig: Reclam Verlag, 1993), pp. 225–42.
63 Van Zundert, 'Author, Editor, Engineer'.

appeal and high risk, and that offers little true insight in the workings of the code hidden behind the interface. Both developers and users of code in the humanities should, therefore, wish to explore different forms of interfacing with code that explicitly acknowledges the interdependence of code and text. We already find ideas on such interdependence of literacies in Knuth.[64] Long dormant, these ideas have now been revived in mixed mode technologies like Jupyter Notebook,[65] which mingle code and text so that the text narrative may elucidate the code narrative. These technologies readily provide affordances for the mixed code and text interfacing that can support the involved evaluation of DH code and scholarship.

We believe that an important step in illuminating the process and results of DH programmers' codework is to develop and explicate reflexive insights into its key epistemological, methodological, and technical aspects. Explaining, for instance, what kind of research questions give impetus to one's codework and how new research insights co-evolve during code development can help both DH programmers and their traditionally trained colleagues recognise the important epistemological connections between humanistic theory and scholarly programming. As a potential starting point in developing such reflections, we provide a set of questions that guided our autoethnographic observations (Appendix 6.A). These questions are not intended to be comprehensive nor prescriptive, but rather aim to initiate a dialogue in the humanities community. In the manner of open-source software, we invite readers to explore, change, and distribute these questions in ways that best suit their research needs.

Our second set of recommendations concerns humanists who do not engage in coding. To start with, it is important to recall that first order logic (i.e. the foundation of most programming languages) has a history that starts with classical philosophy and logic, and inspired the mathematical debates that lead to the formal logic of computer languages. The philosophical roots and history of science, including humanities and computer science, from Aristotle to Bertrand Russell, demonstrate to us a productive interaction between different epistemologies, including code, embodied in the work of scholarly

---

64   Knuth, 'Literate Programming'.
65   Project Jupyter, *Jupyter* (2017), http://jupyter.org/

hybrids, and further advise us that we cannot relegate code to some sort of 'other' style of thought that is foreign to the humanities.

Building expertise to support digital scholarship in the humanities needs a comprehensive framework encompassing epistemological, methodological, technical, and socio-cultural aspects of digital knowledge production. These include developing an understanding of data and code, fostering critical reflection on digital objects of inquiry, and comprehending the influence of algorithmic processes on humanistic investigations. Similarly, training in digital methods should include systematic deliberation on the methodological decisions that influence research processes and results, epistemological and ethical challenges of digital scholarship, how to choose the digital tools and methods that are best suited for specific research questions, and so forth. In our view, we must reach a critical mass of humanities scholars who feel that they have the capacity to understand, if not every line of code, then at least the general thrust of what it is doing and what assumptions it is making.

When it comes to senior scholars, the results of our previous research has shown that humanists favour, and best learn, in practice, when instruction closely follows their area of study, and when it unfolds organically through collaboration with colleagues and students.[66] This is where we see rich potential for developing competencies in digital scholarship among senior humanists as well as early-career researchers — not by trying to turn them into programmers, but by tutoring them in the use of scripts and computer programs while they are engaged in their own practice of scholarship. This means showing the applicability and working of the code in a hands-on way, and qualifying them to provide informed, rather than methodologically myopic feedback on its epistemological qualities. High quality epistemological feedback is, in turn, needed to drive the development of scholarly software code forward. This requires sincere and engaged interaction between scholars who use code, and scholars who produce code: the former must be prepared to treat code and coding as more than questionably relevant; the latter must be prepared to account for their code in an academically recognisable form, such as peer review.

As previously mentioned, codework is necessarily shaped by its social context, which may positively or negatively influence the attitude

---

66 Cf. Antonijević, *Amongst Digital Humanists*.

and perception that both coders and other scholars hold towards their work. Our final set of recommendations thus addresses institutions and organisations, who are best placed to provide the impetus and infrastructure necessary to effect real change in how codework is received in the humanities. A necessary step in that direction is recognition and reward for peer-reviewed digital outputs, including code, as research outputs.[67] A precondition for this is to start grassroots procedures for the peer review of code,[68] and to regard code as alternative expressions of research or epistemologies with equal research value and validity, instead of subordinating code and codework to humanities proper.[69]

Another necessary step is to clarify the nature of the collaboration between coding and non-coding scholars, especially those working on the same project. Too often, DH programmers are treated as service providers instead of research focused scholars, which results in a number of negative consequences. One such consequence is that non-coding humanists appropriate all the effort and results as their own work and invention, even in cases where their only contribution was to provide a question in the form of 'can you do X...?'. Where they contribute substantial research effort and results, DH programmers should be seen as crucial peer collaborators whose competencies create a necessary link between the different areas of expertise. DH programmers have an important responsibility here too in expressing a clear appropriation of, and accountability for, their scientific programming work. Institutions can support such accountability by making it a requirement for publications that involve substantial computational analysis to have clear and comprehensive methodological descriptions of the computational approaches.

---

67 Cf. Nowviskie, *Where Credit is Due*; Todd Presner, 'How to Evaluate Digital Scholarship', *Journal of Digital Humanities*, 1.4 (2012), http://journalofdigitalhumanities.org/1-4/how-to-evaluate-digital-scholarship-by-todd-presner/; American Historical Association, 'Guidelines for the Professional Evaluation of Digital Scholarship by Historians', *American Historical Association* (2015), https://www.historians.org/teaching-and-learning/digital-history-resources/evaluation-of-digital-scholarship-in-history/guidelines-for-the-professional-evaluation-of-digital-scholarship-by-historians (see especially p. 1).
68 Cf. Fitzpatrick, 'Peer Review'.
69 Cf. Burgess and Hamming, 'New Media'; Ramsay and Rockwell, 'Developing Things'.

Appendix 6.A: Survey Questions

1. How does your process of DH programming usually start (e.g., with a research question you want to address; with the data you collected and need to analyse, etc.)? Give examples, if possible.

2. Who is the user you typically have in mind when programming (yourself, your team, the broader DH community)?

3. In what ways, if any, do humanities/digital humanities methods and theories influence your programming? Does this influence differ across the programming phases? Please explain and illustrate.

4. What are the main DH programming decisions you usually need to make? Do you typically think about these decisions in advance or as they appear in practice? Please explain and illustrate.

5. What would be an example of a DH argument you made through programming?

6. What are the main differences and similarities between the arguments you make in DH programming and in DH academic writing?

7. In what ways do you think humanities epistemological and methodological assumptions get reflected in your code?

8. What are the main challenges you experience in DH programming?

9. Is your DH programming typically individual or part of a collaborative project? In what ways, if any, does the decision-making differ in collaborative projects?

10. How and with whom do you usually share your code?

Bibliography

American Historical Association, 'Guidelines for the Professional Evaluation of Digital Scholarship by Historians', *American Historical Association* (2015), https://www.historians.org/teaching-and-learning/digital-history-resources/evaluation-of-digital-scholarship-in-history/guidelines-for-the-professional-evaluation-of-digital-scholarship-by-historians

Anderson, Leon, 'Analytic Autoethnography', *Journal of Contemporary Ethnography*, 35 (2006), 373–95, https://doi.org/10.1177/0891241605280449

Antonijević, Smiljana, *Amongst Digital Humanists: An Ethnographic Study of Digital Knowledge Production* (London, New York: Palgrave Macmillan, 2015), https://doi.org/10.1057/9781137484185

Bauer, Jean, 'Who You Calling Untheoretical?', *Journal of Digital Humanities*, 1 (2011), http://journalofdigitalhumanities.org/1-1/who-you-calling-untheoretical-by-jean-bauer/

Bernardi, Chiara, 'Working Towards a Definition of the Philosophy of Software', *Computational Culture*, 2 (2012), http://computationalculture.net/review/working-towards-a-definition-of-the-philosophy-of-software

Berry, David M., *Critical Theory and the Digital*, Critical Theory and Contemporary Society (New York, NY: Bloomsbury Academic, 2014), https://doi.org/10.5040/9781501302114

—— *The Philosophy of Software: Code and Mediation in the Digital Age* (Basingstoke: Palgrave Macmillan, 2011).

Bezroukov, Nikolai, 'Open Source Software Development as a Special Type of Academic Research: Critique of Vulgar Raymondism', *First Monday*, 4.10 (1999), https://doi.org/10.5210/fm.v4i10.696

Borgman, Christine, 'The Digital Future Is Now: A Call to Action for the Humanities', *Digital Humanities Quarterly*, 3.4 (2009), www.digitalhumanities.org/dhq/vol/3/4/000077/000077.html

Burgess, Helen J., and Jeanne Hamming, 'New Media in Academy: Labor and the Production of Knowledge in Scholarly Multimedia', *Digital Humanities Quarterly*, 5.3 (2011), http://digitalhumanities.org/dhq/vol/5/3/000102/000102.html

Cerquiglini, Bernard, *In Praise of the Variant: A Critical History of Philology* (Baltimore, MD: The Johns Hopkins University Press, 1999).

Coleman, E. Gabriella, *Coding Freedom: The Ethics and Aesthetics of Hacking* (Princeton (US), Woodstock (UK): Princeton University Press, 2013), http://gabriellacoleman.org/Coleman-Coding-Freedom.pdf

Coleman, G., *Hacker, Hoaxer, Whistleblower, Spy: The Many Faces of Anonymous* (London, New York: Verso, 2014).

Coyne, Richard, *Designing Information Technology in the Postmodern Age: From Method to Metaphor*, A Leonardo Book (Cambridge, MA: MIT Press, 1995).

Crombie, Alistair Cameron, *Styles of Scientific Thinking in the European Tradition: The History of Argument and Explanation Especially in the Mathematical and Biomedical Sciences and Arts* (London: Duckworth, 1995).

DeRose, Steven J., et al., 'What Is Text, Really?', *Journal of Computing in Higher Education*, 1 (1990), 3–26, https://doi.org/10.1145/264842.264847

Drucker, Johanna, 'Graphesis: Visual Knowledge Production and Representation', *Poetess Archive Journal*, 2.1 (2010), https://journals.tdl.org/paj/index.php/paj/article/download/4/50

—— 'Humanities Approaches to Graphical Display', *Digital Humanities Quarterly*, 5.1 (2011), http://digitalhumanities.org/dhq/vol/5/1/000091/000091.html

—— 'Should Humanists Visualize Knowledge?', *Vimeo*, video lecture at Lehigh University, Bethlehem, Pennsylvania, 2016, https://vimeo.com/140307034

Fitzpatrick, Kathleen, 'Peer Review, Judgment, and Reading', *Profession* (2011), 196–201, https://doi.org/prof.2011.2011.1.196

Flanders, Julia, and Fotis Jannidis, *Knowledge Organization and Data Modeling in the Humanities* (2015), http://www.wwp.northeastern.edu/outreach/conference/kodm2012/flanders_jannidis_datamodeling.pdf

Forsythe, D., and D. J. Hess, *Studying Those Who Study Us: An Anthropologist in the World of Artificial Intelligence* (Stanford, CA: Stanford University Press, 2001).

Galey, Alan, and Stan Ruecker, 'How a Prototype Argues', *Literary and Linguistic Computing*, 25.4 (2010), 405–24, https://doi.org/10.1093/llc/fqq021

Gurak, Laura, and Smiljana Antonijević, 'Digital Rhetoric and Public Discourse', in *The Sage Handbook of Rhetorical Studies*, ed. by Andrea A. Lunsford, Rosa A. Eberly, and Kirt H. Wilson (London, Thousand Oaks: SAGE Publications, Inc., 2017), pp. 497–508, https://doi.org/10.4135/9781412982795.n26

Hiller, Moritz, 'Signs o' the Times: The Software of Philology and a Philology of Software', *Digital Culture and Society*, 1.1 (2015), 152–63, https://doi.org/10.14361/dcs-2015-0110

Jackson, Virginia, and Lisa Gitelman, 'Introduction', in *'Raw Data' Is an Oxymoron*, ed. by Lisa Gitelman, Geoffrey C. Bowker, and Paul N. Edwards (Cambridge, MA: MIT Press, 2013), pp. 1–14, https://doi.org/10.7551/mitpress/9302.003.0002

Jones, Steven E., *The Emergence of the Digital Humanities* (New York, NY: Routledge, 2014).

Kaltenbrunner, Wolfgang, *Reflexive Inertia: Reinventing Scholarship Through Digital Practices* (Leiden: Leiden University, 2015).

Kemman, Max, Martijn Kleppe, and Stef Scagliola, 'Just Google It: Digital Research Practices of Humanities Scholars', in *Proceedings of the Digital Humanities Congress 2012*, ed. by Clare Mills, Michael Pidd, and Esther Ward (Sheffield: HRI Online Publications, 2014), arXiv:1309.2434, https://www.hrionline.ac.uk/openbook/chapter/dhc2012-kemman

Kirschenbaum, Matthew G., *Mechanisms: New Media and the Forensic Imagination* (Cambridge, MA: MIT Press, 2008).

Kittler, Friedrich, 'Es gibt keine Software', in *Draculas Vermächtmis* (Leipzig: Reclam Verlag, 1993), pp. 225–42.

Knuth, Donald E., 'Literate Programming', *The Computer Journal*, 27.2 (1984), 97–111, https://doi.org/10.1093/comjnl/27.2.97

Lassiter, L. E., *The Chicago Guide to Collaborative Ethnography*, Chicago Guides to Writing, Editing and Publishing (Chicago, London: University of Chicago Press, 2005), http://bit.ly/2iLCmGY

Latour, Bruno, *Science in Action: How to Follow Scientists and Engineers Through Society* (Cambridge, MA: Harvard University Press, 1988).

—— 'Where Are the Missing Masses, Sociology of a Few Mundane Artefacts', in *Shaping Technology-Building Society. Studies in Sociotechnical Change*, ed. by Wiebe Bijker and John Law (Cambridge, MA: MIT Press, 1992), pp. 225–59, http://www.bruno-latour.fr/node/258

Liu, Alan, 'Digital Humanities and Academic Change', *English Language Notes*, 47.1 (2009), 17–35, https://doi.org/10.1215/00138282-47.1.17

Mackenzie, Adrian, 'The Performativity of Code', *Theory, Culture & Society*, 22.1 (2005), 71–92, https://doi.org/10.1177/0263276405048436

Manovich, Lev, *Software Takes Command: Extending the Language of New Media*, International Texts in Critical Media Aesthetics 5 (New York, NY: Bloomsbury Academic, 2013).

Marino, Mark C., 'Field Report for Critical Code Studies, 2014', *Computational Culture*, 4 (2014), http://computationalculture.net/article/field-report-for-critical-code-studies-2014%E2%80%A8

McCarty, Willard, *Humanities Computing* (Basingstoke: Palgrave Macmillan, 2005).

McDowell, Charlie, et al., 'The Impact of Pair Programming on Student Performance, Perception and Persistence', in *Proceedings of the 25th International Conference on Software Engineering*, ICSE '03 (Washington, DC: IEEE Computer Society, 2003), pp. 602–07, http://dl.acm.org/citation.cfm?id=776816.776899

McPherson, Tara, 'Why Are the Digital Humanities So White? Or Thinking the Histories of Race and Computation', in *Debates in the Digital Humanities*, ed. by Matthew K. Gold (Minneapolis: University of Minnesota Press, 2012), pp.

139–60, https://doi.org/10.5749/minnesota/9780816677948.003.0017, http://dhdebates.gc.cuny.edu/debates/text/29

Munson, M. Arthur, 'A Study on the Importance of and Time Spent on Different Modeling Steps', *SIGKDD Explorations*, 13 (2011), 65–71, https://doi.org/10.1145/2207243.2207253

Nader, Laura, 'Up the Anthropologist: Perspectives Gained from Studying Up', in *Reinventing Anthropology*, ed. by D. H. Hymes, Ann Arbor Paperbacks Series (University of Michigan Press, 1972), pp. 284–311, http://www.dourish.com/classes/readings/Nader-StudyingUp.pdf

Nowviskie, Bethany, ed., *#Alt-Academy 01: Alternative Academic Careers for Humanities Scholars* (New York: MediaCommons Press, 2014), http://mediacommons.org/alt-ac/

—— 'Where Credit Is Due: Preconditions for the Evaluation of Collaborative Digital Scholarship', *Profession* (2011), 169–81, https://doi.org/10.1632/prof.2011.2011.1.169

Presner, Todd, 'How to Evaluate Digital Scholarship', *Journal of Digital Humanities*, 1.4 (2012), http://journalofdigitalhumanities.org/1-4/how-to-evaluate-digital-scholarship-by-todd-presner/

Project Jupyter, *Jupyter* (2017), http://jupyter.org/

Ramsay, Stephen, and Geoffrey Rockwell, 'Developing Things: Notes toward an Epistemology of Building in the Digital Humanities', in *Debates in the Digital Humanities*, ed. by Matthew K. Gold (Minneapolis: University of Minnesota Press, 2012), pp. 75–84, https://doi.org/10.5749/minnesota/9780816677948.003.0010, http://dhdebates.gc.cuny.edu/debates/text/11

Rockwell, Geoffrey, 'What Is Text Analysis, Really?', *Literary and Linguistic Computing*, 18.2 (2003), 209–19, https://doi.org/10.1093/llc/18.2.209

Rossum, Guido van, Barry Warsaw, and Nick Coghlan, 'PEP 8 — Style Guide for Python Code', *Python* (5 July 2001), https://www.python.org/dev/peps/pep-0008/

Schreibman, Susan, and Ann M. Hanlon, 'Determining Value for Digital Humanities Tools: Report on a Survey of Tool Developers', *Digital Humanities Quarterly*, 4.2 (2010), http://digitalhumanities.org/dhq/vol/4/2/000083/000083.html

Schreibman, Susan, Laura Mandell, and Stephen Olsen, 'Introduction', *Profession* (2011), 123–201, https://doi.org/10.1632/prof.2011.2011.1.123

Smith, James, 'Coding and Digital Humanities', *James Gottlieb: Seeing What Happens When You Collide the Humanities with the Digital* (8 March 2012), http://www.jamesgottlieb.com/old/2012/03/coding-and-digital-humanities/

Spaaks, Juriaan H., 'The Research Software Directory and How It Promotes Software Citation: Improve the Findability, Citability, and Reproducibility of Research Software', *EScience Center* (11 December 2018), https://blog.esciencecenter.nl/the-research-software-directory-and-how-it-promotes-software-citation-4bd2137a6b8

Svensson, P., *Big Digital Humanities: Imagining a Meeting Place for the Humanities and the Digital* (Digital Culture Books, Ann Arbor, MI: University of Michigan Press, 2016), https://doi.org/10.1353/book.52252, http://hdl.handle.net/2027/spo.13607060.0001.001

Takats, Sean, 'A Digital Humanities Tenure Case, Part 2: Letters and Committees', *The Quintessence of Ham* (7 February 2013), http://quintessenceofham.org/2013/02/07/a-digital-humanities-tenure-case-part-2-letters-and-committees/

Williams, Peter, et al., 'The Role and Future of the Monograph in Arts and Humanities Research', *Aslib Proceedings*, 61.1 (2009), 67–82, https://doi.org/10.1108/00012530910932294

Zundert, Joris J. van, 'Author, Editor, Engineer: Code & the Rewriting of Authorship in Scholarly Editing', *Interdisciplinary Science Reviews*, 40.4 (2016), 349–75, https://doi.org/10.1080/03080188.2016.1165453

Zundert, Joris J. van, and Ronald Haentjens Dekker, 'Code, Scholarship, and Criticism: When Is Coding Scholarship and When Is It Not?', *Digital Scholarship in the Humanities* (2017), https://doi.org/10.1093/llc/fqx006

7. The Evaluation and Peer Review of Digital Scholarship in the Humanities

Experiences, Discussions, and Histories

Julianne Nyhan

Introduction

The project of publishing guidelines and advocacy documents for the evaluation of digital scholarship in the humanities has gained particular momentum since *c.* 2002. This 'turn' is unlikely to have been spontaneous, and thus various questions follow: which contexts and what interests shaped the work of devising guidelines for the evaluation of digital scholarship? What were the digital humanities communities' experiences of the evaluation of digital scholarship during the years before *c.* 2002? And what trajectory has the evaluation of digital scholarship followed over the longer term? In short: what is the history of the take-up and development of evaluative methods for the assessment of digital scholarship in the humanities? In this chapter, I explore these wider questions by looking more closely at how the evaluation of digital scholarship was experienced and discussed by the humanities computing community during the years before *c.* 2002. This chapter contributes to this volume by presenting an overview of the trajectory and contours of the debates about digital scholarship and communication that occurred in the humanities computing community. Chronologically 'downstream' of the digital humanities, the material presented in this chapter offers useful and grounded preliminary and historical material that explains some of the longer-term origins of many of the debates that still concern the digital humanities, which are discussed in the introduction to this volume in particular, but in other chapters too.

Digital humanities is often said to have developed from humanities computing, whose origins, in turn, are often traced to approximately 1949.[1] As will be shown below, conversations about the evaluation of the field's digital scholarship, as well as a few projects that sought to tackle its various aspects, can be documented from at least the 1960s. Yet, it is in the first decade of the twenty-first century that a cluster of publications and projects about evaluation can be noted, many of them influential. In 2002, the MLA (Modern Language Association) Committee on Information Technology published 'Guidelines for Evaluating Work in Digital Humanities and Digital Media'.[2] These guidelines have proved to be a significant starting point for those seeking direction about the evaluation of digital scholarship.[3] In 2004, the Networked Infrastructure for Nineteenth-Century Electronic Scholarship (NINES) was set up with aims that included its functioning as a peer review collective for digital work about the nineteenth century.[4] The work on evaluation conducted by Geoffrey Rockwell from 2005 to 2008 was officially released by the MLA's Committee on Information Technology in 2008.[5] New peer-reviewed platforms for the digital publication of multimedia scholarship (for example, *Vectors*)

---

1 See, for example, John Unsworth, *Digital Humanities Beyond Representation* (Orlando, FL: University of Central Florida, 2006), http://www.people.virginia.edu/~jmu2m/UCF/

2 Modern Language Association of America, 'Guidelines for Evaluating Work in Digital Humanities and Digital Media', *Modern Language Association* (2012), https://www.mla.org/About-Us/Governance/Committees/Committee-Listings/Professional-Issues/Committee-on-Information-Technology/Guidelines-for-Evaluating-Work-in-Digital-Humanities-and-Digital-Media

3 See, for example, Geoffrey Rockwell, 'On the Evaluation of Digital Media as Scholarship', *Profession* (2011), 152–68, https://doi.org/10.1632/prof.2011.2011.1.152

4 Jerome McGann, 'On Creating a Usable Future', *Profession* (2011), 182–95, https://doi.org/10.1632/prof.2011.2011.1.182. Notable precursors include the collective that was set up in 1998 by Suda online (SOL), which included an innovative form of online peer review of the translations and annotations made to it by users. See Raphael Finkel et al., 'The Suda On Line (www.stoa.org/sol/)', *Syllecta Classica*, 11 (2000), 178–90, https://doi.org/10.1353/syl.2000.0005

5 Susan Schreibman, Laura Mandell, and Stephen Olsen, 'Introduction', *Profession* (2011), 123–201 (p. 127), https://doi.org/10.1632/prof.2011.2011.1.123

began publishing in 2005.⁶ Around this time, the MLA's Committee on Scholarly Editions incorporated electronic editions into its guidelines for print editions.⁷ In 2006, the MLA also stated that '[d]epartments and institutions should recognize the legitimacy of scholarship produced in new media, whether by individuals or in collaboration, and create procedures for evaluating these forms of scholarship'.⁸ That same year the influential ACLS (American Council of Learned Societies) Commission on Cyberinfrastructure in the Humanities and Social Sciences also emphasised the importance of recognising digital scholarship, including evaluating it appropriately.⁹ In 2007, the report 'University Publishing in a Digital Age' urged universities to show 'a renewed commitment to publishing in its broadest sense'.¹⁰

The documents and projects outlined above are, *ceteris paribus*, in favour of digital scholarship and committed to devising robust ways of assessing it. Yet, regarding the 2006 quotation above from the MLA (about the worth of digital scholarship and the necessity of devising approaches to its assessment), the fact that it was necessary to make such a statement implies that the reception and evaluation of digital scholarship remained problematic. On my initial reading of the documents cited above, given their emphasis on the necessity for evaluating and recognising digital scholarship, I assumed that the imagined audience for such calls was the wider academy. Yet, I began to wonder about attitudes to, and experiences of, evaluation that may have existed in the humanities computing community itself. Was the

---

6 Tara McPherson, 'Scaling Vectors: Thoughts on the Future of Scholarly Communication', *Journal of Electronic Publishing*, 13.2 (2010), https://doi.org/10.3998/3336451.0013.208

7 See Modern Languages Association of America Task Force for Evaluating Scholarship for Tenure and Promotion, *Report of the MLA Task Force on Evaluating Scholarship for Tenure and Promotion* (New York: MLA, 2006), p. 42, http://www.mla.org/pdf/taskforcereport0608.pdf

8 Modern Languages Association of America Task Force, *Report of the MLA Task Force*, p. 11.

9 American Council of Learned Societies, *Our Cultural Commonwealth: The Report of the American Council of Learned Societies Commission on Cyberinfrastructure for the Humanities and Social Sciences* (New York: American Council of Learned Societies, 2006), p. 34, https://www.acls.org/uploadedFiles/Publications/Programs/Our_Cultural_Commonwealth.pdf

10 Laura Brown, Rebecca Griffiths, and Matthew Rascoff, 'University Publishing in a Digital Age', *The Journal of Electronic Publishing*, 10.3 (2007), https://quod.lib.umich.edu/j/jep/3336451.0010.301?view=text;rgn=main, https://doi.org/10.3998/3336451.0010.301

community united in favour of digital scholarship being formally evaluated? Was there internal agreement about what constituted digital scholarship and appropriate forms of evaluation?

In order to explore these questions further, and thus to understand more about the prehistory of the evaluation of digital scholarship, I will survey some of the conversations the humanities computing community recorded in the years before c. 2002 concerning peer review and evaluation. In particular, I will uncover and discuss attitudes to and experiences of the evaluation of digital, or digitally-derived, research recorded in internet and www forums, publications, and oral history interviews.[11]

Because humanities scholarship is usually evaluated via peer review, I will survey conversations about one or both of these terms. I define the terms 'peer review' and 'evaluation' broadly to include any kind of assessment (whether qualitative or quantitative) of digital scholarship that is discussed in the literature I have surveyed. So too, I have adopted a broad definition of digital scholarship that includes not only digital or digitally-derived scholarship but also scholarship that has been published digitally. I do this on account of the practice of 'double-publication', which has long been at play in the digital humanities, where publication about a digital humanities artefact or tool is required in addition to the digital object or resource itself.[12]

A growing body of literature addresses the evaluation of digital scholarship and the issues connected to it. Important discussions include the social and dialogic contexts that might be cultivated at a departmental level to support the longer-term evaluation of digital scholarship,[13]

---

11   The literature that I surveyed covered the main journals in the field that were published from the setting up of computing and the humanities onwards (*Computing and the Humanities; Literary and Linguistic Computing / DSH: The Journal of Digital Scholarship in the Humanities; Digital Humanities Quarterly; Digital Studies / Le champ numérique / Text Technology / CHWP: Computing in the Humanities Working Papers*). I also surveyed the grey literature that I had access to, namely the transactions of Humanist; the newsletter of the Association for Computers and the Humanities (ACH); early issues of the *ALLC Bulletin*; and online proceedings of the ALLC/ Digital Humanities conferences.

12   'Scholarship in electronic formats seems to be recognized when done in addition to work in print formats but may place a candidate at risk if presented as the sole or primary scholarly basis for consideration for tenure.' Modern Languages Association of America Task Force, *Report*, p. 44.

13   Rockwell, 'On the Evaluation of Digital Media'.

criteria for evaluative committees who assess digital scholarship,[14] and the particular circumstances that often underpin digital scholarship, for example, collaboration.[15] Publications also advocate for the necessity of evaluating digital scholarship,[16] explore ways in which particular communities might contribute to evaluation,[17] and discuss some approaches to assessing emerging forms of digital scholarship.[18] Yet, the wider history of the evaluation and peer review of digital scholarship is little addressed (while the history of peer review in the humanities also requires further research).[19] This paper seeks to explore this topic by sketching the ways in which peer review and evaluation were discussed and understood by the humanities computing community during the years before *c*. 2002.

Experiences and Discussion of Evaluation *c*. 1963–2001

The discussions and debates that are summarised below are founded on the following questions: what constitutes a digital research output? Which outputs should be formally evaluated? In line with what criteria could they be evaluated? How should the peer review process be organised and managed, and who might participate in it? What do bibliometrics imply about the perceived impact and quality of digital scholarship? The responses these questions elicited are often underscored by a certain ambivalence about the robustness and fair-mindedness of the process of evaluating digital scholarship. The question of whether digital scholarship could even get a fair hearing

---

14 Kathleen Fitzpatrick, 'Peer Review, Judgment, and Reading', *Profession* (2011), 196–201, https://doi.org/10.1632/prof.2011.2011.1.196
15 Bethany Nowviskie, 'Where Credit Is Due: Preconditions for the Evaluation of Collaborative Digital Scholarship', *Profession* (2011), 169–81, https://doi.org/10.1632/prof.2011.2011.1.169
16 Schreibman, Mandell, and Olsen, 'Introduction'.
17 Sarah L. Pfannenschmidt and Tanya E. Clement, 'Evaluating Digital Scholarship: Suggestions and Strategies for the Text Encoding Initiative', *Journal of the Text Encoding Initiative* (2014), 7, https://doi.org/10.4000/jtei.949
18 Steve Anderson and Tara McPherson, 'Engaging Digital Scholarship: Thoughts on Evaluating Multimedia Scholarship', *Profession* (2011), 136–51, https://doi.org/10.1632/prof.2011.2011.1.136
19 Noah Moxham and Aileen Fyfe, 'The Royal Society and the Prehistory of Peer Review, 1665–1965', *The Historical Journal The Historical Journal* 61.4 (2018), 863–889, p. 886, https://doi.org/10.1017/S0018246X17000334

seems to be raised implicitly. At the time of writing, digital humanities is apparently in a strong position, so this attitude might seem puzzling to readers of this chapter. Yet, it is an important backdrop against which many of the conversations summarised below should be read, and I will therefore briefly address it and its wider contexts.

Individual and Group Experiences of Making Digital Scholarship

References to the negative evaluations some humanities computing scholars have received of their digital work feature in oral history interviews, listserv discussions, and formal publications. Of course, negative evaluations were not a universal experience, as evidence from Julianne Nyhan and Andrew Flinn's oral history interviews demonstrated: Susan Hockey and John Nitti, for example, recalled the positive collaborations they pursued with established humanities scholars.[20] However, others readily recalled the opposition their work met with. For example, Mary Dee Harris reported that: 'I got a lot of flak from the Department about my work. One of the graduate advisers swore that I was trying to destroy literature by using the computer.'[21] John Burrows and Hugh Craig discussed the difficulties they sometimes faced when trying to publish their scholarship in 'mainstream' English journals, as opposed to dedicated humanities computing or digital humanities publications.[22]

Some discussions on Humanist (which is referred to as an electronic seminar, see the further discussion of it below) tally with these experiences. For instance, a post to Humanist emphasised that there existed almost 'universal disregard for work in computing among the committees that govern hiring, tenure, and promotion'.[23] Another post pointedly asked: 'Do tenure and promotion committees value

---

20 Julianne Nyhan and Andrew Flinn, *Computation and the Humanities: Towards an Oral History of Digital Humanities*, 1st ed. (Cham, Switzerland: Springer, 2016), pp. 87–97, 137–56.
21 Ibid., p. 125.
22 Ibid., p. 49.
23 *Humanist Discussion Group Archive (1987–2018)*, 1.49, ed. by Willard McCarty (1987/88), http://dhhumanist.org/. The archives of Humanist that are cited in this chapter are accessible via the following landing page: http://dhhumanist.org/

programming, software reviewing, and other of the activities [sic] so typical of HUMANIST addressees?'.[24] These sentiments find an echo in formally published literature too. A. Q. Morton, for example, in 1963 recalled how his work was dismissed by the humanities journals he first sought to publish it in:

> The first technical article I wrote I sent to the *Scottish Journal of Theology*. It arrived back within three days. I sent it to the *Expository Times*. A letter came back: 'Dear Mr. Morton, I do not understand this but I am quite sure that if I did understand it, it would be of no value.' I sent it to *Science News*, whose editor came up to see me about immediate publication.[25]

Joel D. Goldfield echoed this experience of dismissal when, in 1993, he wrote approvingly of Paul Fortier's strategy for de-centring his computational techniques and data:

> At this juncture I therefore accept Paul Fortier's politically wise approach in his study on Gide's *L'immoraliste*: statistical sophistication in stylometric and thematic analysis, as well as statistical details implicit in the interpretation, are relegated to appendices or simply not included in the publication.[26]

Indeed, Joseph Raben, who was for many years the editor of *Computers and the Humanities* (the field's first academic journal), indicated the problem was a systemic one. In 1991, he wrote:

> for many individuals the mere existence of this journal [*Computers and the Humanities*] has meant the difference between academic success and failure. Promotion and tenure committees, restricted in their vision to 'legitimate publication' have often been satisfied by articles that have passed our referees and appeared in our pages. Few of these articles would have been appropriate for the conventional journals of their respective disciplines.[27]

Other conversations indicate it was not only the use of a computer for research that was considered problematic by some; merely publishing

---

24   *Humanist*, 1.47 (1987/88).
25   A. Q. Morton, 'A Computer Challenges the Church', *The Observer (1901–2003)* (3 November 1963), p. 21.
26   Joel D. Goldfield, 'An Argument for Single-Author and Similar Studies Using Quantitative Methods: Is There Safety in Numbers?', *Computers and the Humanities*, 27.5–6 (1993), 365–74 (p. 370), https://doi.org/10.1007/BF01829387
27   Joseph Raben, 'Humanities Computing 25 Years Later', *Computers and the Humanities*, 25.6 (1991), 341–50 (p. 341), https://doi.org/10.1007/bf00141184

work on a digital platform could also be viewed as problematic. An example from 1987 speaks to this. The idea of setting up an electronic journal was proposed for the field of humanities computing, with peer review by an editorial board.[28] The idea was rejected on various grounds, including the proposed medium of publication: it was felt that few researchers would contribute to it, as the electronic format held too many risks.[29] Willard McCarty also claimed that electronic publication in the humanities was devoid of 'professional kudos' and had the potential to 'pre-empt [...] conventional [publication]'.[30]

In this way, I believe, the inauspicious reception digital scholarship sometimes received from the wider community partly explains the ambivalence some members of the humanities computing community expressed towards the evaluation of digital scholarship in the conversations summarised below.[31] The conversations that took place about the evaluation of digital scholarship will now be presented, beginning with discussions about which outputs were considered amenable to peer review.

What Should Be Evaluated?

One of the richest sources of discussion about experiences of, and attitudes to, the evaluation and peer review of digital scholarship I have encountered is contained in the archives of Humanist. Humanist was established in 1987 on the BITNET/NetNorth/EARN node in Toronto, Canada, and run on Listserv software.[32] It was styled as an academic seminar, and debates about the evaluation of digital scholarship occurred on it from an early stage. In the earliest Humanist posts, questions about peer review are somewhat inward looking: one question asked was whether a form of peer review, in the sense of moderation, should be

---

28   *Humanist*, 1.44 (1987/88).
29   *Humanist*, 1.49 (1987/88).
30   Willard McCarty, 'Humanist So Far: A Review of the First Two Months', *ACH Newsletter*, 9.3 (1987).
31   Though not within the scope of this article, the numerous debates that have taken place in the wider academy that question peer review are presumably also relevant to this. See, for example, Daryl E. Chubin and Edward J. Hackett, *Peerless Science: Peer Review and U.S. Science Policy* (Albany, NY: SUNY Press, 1990).
32   Willard McCarty, 'HUMANIST: Lessons from a Global Electronic Seminar', *Computers and the Humanities*, 26.3 (1992), 205–22 (p. 205–06), https://doi.org/10.1007/BF00058618

applied to Humanist itself,[33] a proposal that was ultimately rejected.[34] Another question asked whether posts to Humanist might be peer reviewed so they could be counted by tenure committees.[35] Discussions about whether posts to a listserv group might be peer reviewed now seem antithetical to the participatory and interactive paradigm that currently characterises many digitally-mediated communication platforms. Such conversations remind us of the novelty of the technology at that stage, and they prompt questions about how social contexts and dialogue, and not just technological affordances, shaped the take up of computing in the humanities. As we shall see, over the longer term, social and dialogic factors also played a role in persuading the humanities computing community of the necessity of formally evaluating digital scholarship in the humanities.

Conversations on Humanist soon turn to the absence of peer review mechanisms for humanities computing scholarship (including electronically published articles and studies like editions, software, code, tools, and other kinds of computational work and software reviews). In the discussions this observation gives rise to, or interlinks with, ambivalence towards the field of humanities computing itself is palpable. When summarising the first two months of conversations that had taken place on Humanist, McCarty noted that frustration had been expressed with the 'juvenality [sic] of an emerging discipline: the lack of peer review, hence of quality-control'.[36] Indeed, in a post to Humanist, McCarty argued that peer review was essential to reforming the status quo:

> The second reason for the disregard from our academic masters and colleagues may be the often poor quality of the writing (and sometimes thinking) associated with computing. The informality of the medium may have quite a bit to do with this. Mainframe editors are in general so primitive and screen images so difficult to proofread that we are tempted to slap something down and dash it off without much thought. We can do something about this, it has been suggested, by peer-review and editorial intervention.[37]

---

33 See, for example, *Humanist* 1.28 (1987/88).
34 McCarty, 'HUMANIST: Lessons', 210–12.
35 *Humanist*, 1.40 (1987/88).
36 McCarty, 'Humanist So Far', p. 2.
37 *Humanist*, 1. 49 (1987/8).

Responses to McCarty were mixed; over the longer term, doubts about the imprimatur of peer review continued to be raised. The question of how evaluation intersected with disciplinary identity was evoked when peer review was discussed as being a hallmark of the established humanities, a sector from which humanities computing tended to differentiate itself: 'if we really boil things down to their foundations and meanings, we may find that a lot of them are rubbish and that the Mainstream with its Peer Reviewers is largely unsatisfactory'.[38] All the same, a tentative acceptance of the necessity of some form of peer review or formal evaluation of the field's scholarship is indicated by some. For example, by 1996/7, a contentious critique of the Text Encoding Initiative (TEI) Guidelines observed that 'the TEI Guidelines have never been subjected to significant peer review'.[39] Whatever the accuracy of the claim, that the guidelines should be criticised in such terms implies that peer review was seen as increasingly fundamental.

Which Evaluative Criteria?

Various concerns were also raised about the difficulties of actually implementing peer review. It was recognised that, in order to elaborate peer review guidelines, complex, fundamental, and likely contested questions about what constituted quality would have to be addressed. For example: 'Both Charles Faulhaber and Willard McCarty imply that peer review is enough to put e-work on an equal footing with conventional work. But are there any criteria for peer review? [...] without some rules, isn't it a meaningless criterion?'.[40]

Reaching a consensus about how quality could be identified was just part of the task. Identifying those with the technical skills necessary to evaluate such work was also germane, as was the ongoing problem of what could and should count as scholarship:

> The problem is that none of the people in my department would be able to judge work in computers, since they use the computer mostly as a typewriter, with some network involvement. Not to badmouth my own department, this would be true of most departments I know of.

---

38 *Humanist*, 14.52 (2000).
39 *Humanist*, 10.789 (1996/7).
40 *Humanist*, 12.1040 (1998/9); see also 1.344 (1987/8).

[…] Next, there is the problem of who does the work. I know of people who have published concordances, for example, who downloaded the text, outsourced the programming, made a KWIC concordance, so there was little formatting, got it published and submitted it to the tenure committee. Such work should not count. On the other hand, if you write a concordance program yourself, no matter how good, you will have a hard time getting any credit for it.[41]

Organising the Peer Review Process

The conversations on Humanist range over various possibilities for how peer review could be organised and implemented. Overall, one is struck by the conservative nature of these posts. It is curious to see fairly standard humanities approaches being mooted as viable approaches to assessing scholarship that often did not fit into the pre-determined categories of the mainstream humanities. The old chestnut of appointing a group of esteemed scholars to devise evaluative guidelines was proposed:

> A procedure should be established by professional organizations and the e-text center for the peer-reviewing of annotated e-texts if they are tagged beyond screen mark-up (e.g., morphological and literary tagging). This reviewing could take place prior to or following in-house editing, depending on the expertise of the reviewers.[42]

A post about how peer review could be applied to a pre-publication initiative suggested that:

> an editorial board, as prestigious as possible could be organized and could begin selecting the better papers so as to provide a quality of intellectual certification through some classical peer review […] The selected papers could then be marked in such a way that users would know that they are fully certified as if they had been published in a normal, peer-reviewed journal.[43]

Some posts did consider a more innovative form of peer review that could potentially subvert established hierarchies:

> Could the use of […] 'e-review' methods eventually supplant the existing system of peer review used by conventional publishers (the lack of which

---

41   *Humanist*, 12.1050 (1999).
42   *Humanist*, 5.881 (2085) (1991/2).
43   *Humanist*, 13.221 (221) (1999/2000).

is one of the reasons libraries are reluctant to buy self-published books)? Are there any other Humanists out there who have experimented in self-publishing?[44]

A more differentiated approach to peer review was also suggested:

> it seems clear that the user needs to know whether what he or she has on screen is worth spending time puzzling over. Peer-review seems to me essential for some kinds of online publication (journals, usw.), but not for everything. Given a disciplined self, self-publication can be (a) a powerful inducement for our colleagues to get involved, and (b) a way of getting into the public light interesting, valuable material that otherwise would stay in darkness. The more conversational (like Humanist), the more experimental the less peer-review seems appropriate.[45]

The question of how and why it was that some processes went on to be largely adopted by the field is a question that remains open for further studies in this area to explore.

Implicit Peer Review

A prominent debate that played out on Humanist, and continued in *Computers and the Humanities,* again showed the complicated relationship the field of humanities computing had with evaluation and peer review. In 1992, Mark Olsen criticised humanities computing for its 'intellectual failure', as evidenced by the implicit and explicit peer review of its work:

> Our failure is indicated by both explicit and implicit peer review of our work. Implicitly by the intellectual failure of humanities computing research to be cited by or published in (with a few notable exceptions) mainstream scholarship. Bluntly put, scholars in our home disciplines (literature, history, etc.) seem to be able to safely ignore the considerable literature generated by humanities computing research over the years. Explicit peer review is indicated, in part, by the fact that humanities computing hasn't been invited to the banquet. We don't *have* to be invited precisely because the results of so much work can be ignored by scholarship in our home disciplines.[46]

The following year, he published a more detailed version of this argument in a special edition of *Computers and the Humanities,* together with a set

---

44 *Humanist,* 7.453 (836) (1993/4).
45 *Humanist,* 9.872 (916) (1995/6).
46 *Humanist,* 6.652 (845) (1992/3).

of responses from the wider humanities computing community. Olsen wrote how his argument had caused 'considerable debate concerning the proper methods of disciplinary evaluation',[47] and again emphasised the importance of peer review, including the notion of implicit peer review and what it said about the field:

> Given the dominance of peer review in scientific and humanities research, as demonstrated in publication evaluation, grant applications, and hiring/tenure decisions, I find it very difficult to discount the importance of the most objective measure of the value of our work to our peers the decision to read, to use, and to publish our conclusions.[48]

Goldfield's response to Olsen acknowledges humanities computing's marginalisation, but he nonetheless detects the advent of 'a long-awaited, but still incipient, success *d'etre enfin parvenus*'.[49] Arguing that the field was 'battling on two fronts, one scholarly and one political',[50] he discusses its ambivalent attitude towards the peer review of digital scholarship:

> I find fallacious [Olsen's] implicit assumption that studies of interest, new truths, and allegations quickly find their way into the mainstream in the humanities. I would submit that there are two compelling factors working against mainstream entry and fertilization in our quantitative interdiscipline. The first is the inertia of mainstream journals' reviewers and possibly editors, and the unwillingness of the studies' authors to submit their work for peer review, especially in a form palatable for the keepers of the keys.[51]

Nevertheless, during the years under discussion various peer review initiatives were undertaken. For example, the *ACH Newsletter* includes a notice that IBM had funded the MLA and the 'Center for Applied Linguistics to implement a system of peer review for language-oriented software written for IBM microcomputers and compatible hardware'.[52] Yet, the impact of such initiatives on the humanities computing

---

47 Mark Olsen, 'Critical Theory and Textual Computing: Comments and Suggestions', *Computers and the Humanities*, 27.5–6 (1993), 395–400 (p. 395), https://doi.org/10.1007/BF01829390
48 Olsen, 'Critical Theory', 395–96.
49 Goldfield, 'An Argument for Single-Author', 371.
50 Ibid., 366.
51 Ibid., 371.
52 'IBM Grants', *ACH Newsletter*, 9.3 (1987), p. 6.

community appears to have been limited. Six years later the lack of progress made in the context of peer review was again addressed, and the community was once more reminded that 'the production of peer reviewed scholarship is the single most important activity for professional advancement in academe, including tenure, promotion, and salary increases'.[53]

From the late 1990s onwards, there are notable signs that the rejection of the digital per se was coming to an end. One contributor to Humanist wrote of developments at UC Berkeley:

> I have finally gotten my hands on the formal statement proposed by Berkeley's Library Committee to the campus's Academic Senate, with respe[c]t to faculty review and different media: 'In the course of reviewing faculty for merit and promotion, when there are grounds for believing that processes of peer review and quality assurance are the same in different media, equal value should be attached to the different forms of scholarly communication'.[54]

Other notable developments include the announcement of a new electronic imprint from the University of Virginia Press, and its intention to

> look nationally and internationally for pioneering digital work that emphasizes both creative scholarship and innovative technology. Each project published will be approved by the press's editorial board and will receive extensive peer review just as print publications do.[55]

In 2002, an essay 'recently published by the Knight Higher Education Collaborative [argued that] universities and colleges should establish policies declaring peer-reviewed work in electronic form suitable for consideration in promotion and tenure decisions'.[56] Nevertheless, the essay noted that some scholars still needed reassurance that electronic publication would not harm their careers.[57]

---

53 Stéfan Sinclair et al., 'Peer Review of Humanities Computing Software', in *ALLC/ACH 2003 — Conference Abstracts*, ([n.p.], 2003), pp. 143–45.
54 *Humanist*, 13.72 (1999/2000).
55 *Humanist*, 15.524 (2001/2).
56 *Humanist*, 15.724 (2001/2).
57 Ibid.

Conclusion

The material cited above shows that many fundamental conversations took place in the years before *c.* 2002 in the humanities computing community about what constituted academic and technical excellence in digital and digitally-derived scholarship, about the appropriateness of peer review as a mechanism for evaluating digital scholarship, and about whether the digital was a suitable medium for publication. On the whole, the evidence I have gathered here suggests the community had mixed experiences of, and attitudes toward, peer review and formal evaluation. While a consensus does seem to have been reached about the importance of formal evaluation for the emerging discipline, this review indicates that it took time to build such a consensus (and, of course, agreement was not necessarily unanimous). Discussion and debate seem to have played a crucial role in building this consensus over the longer term.

External factors, such as the growing acceptance of digital publication, may also have offered the community an important signal that change was on the horizon and they would need to respond accordingly. It also seems reasonable to propose that the wider position of digital humanities, which by *c.* 2002 was undergoing a process of institutionalisation, made the requirement for evaluative guidelines all the more urgent.[58] Indeed, Matthew G. Kirschenbaum has noted a 'rapid and remarkable rise'[59] of the term 'digital humanities' around this time. He has written of the 'surprisingly specific circumstances'[60] that arguably led to the rise of the term, and that included the preparations

---

58 By 2013, Matthew L. Jockers, for example, discussed the rapidly institutionalising field thus: 'Academic jobs for candidates with expertise in the intersection between the humanities and technology are becoming more and more common, and a younger constituent of digital natives is quickly overtaking the aging elders of the tribe. [...] Especially impressive has been the news from Canada. Almost all of the "G10" (that is, the top thirteen research institutions of Canada) have institutionalized digital humanities activities in the form of degrees [...] programs [...] or through institutes [...]'. Matthew L. Jockers, *Macroanalysis: Digital Methods and Literary History*, 1st ed. (Urbana: University of Illinois Press, 2013), pp. 13–14, https://doi.org/10.5406/illinois/9780252037528.001.0001

59 Matthew G. Kirschenbaum, 'What Is Digital Humanities and What's It Doing in English Departments?', *ADE Bulletin* (2010), 55–61 (p. 56) https://doi.org/10.1632/ade.150.55

60 Kirschenbaum, 'What is Digital Humanities', 56.

(from c. 2001 until its publication in 2004) of Blackwell's *Companion to Digital Humanities*, the establishment of the Alliance of Digital Humanities Organizations (ADHO) in 2005, and the establishment of the Digital Humanities initiative by the NEH (National Endowment for the Humanities) in 2006 (which became the Office of Digital Humanities in 2008).[61] He wrote that '[i]n the space of a little more than five years, digital humanities had gone from being a term of convenience used by a group of researchers who had already been working together for years to something like a movement'.[62] Advances in the digital evaluation of scholarship, such as I have discussed above, are not included in Kirschenbaum's list. Is it merely a coincidence that peer review efforts bear a particular kind of fruit, and exert a specific influence, around the time of the 'rise' of the term digital humanities? Is it plausible to suggest that progress made in the digital evaluation of scholarship contributed to the institutionalisation of the digital humanities? And, if that is the case, what role might digital evaluation play in the ongoing development and institutionalisation of the digital humanities? These are questions that subsequent research about the history of peer review and evaluation of digital scholarship might take up.

The institutionalisation of the digital humanities is *in media res*. Much progress has been made in important areas like faculty appointments, the establishment of dedicated teaching programmes, and the setting up of prestigious centres.[63] Nevertheless, much remains to be done to address ongoing questions that are pertinent to securing a firmer foothold, including, for example, urgent work on areas like the epistemology of the digital (such as appears in chapters 3 and 6 of this volume), and in terms of analysing and theorising the multi-layered and sometimes tacit scholarship that informs and is embodied in the computational artefacts the field creates.[64] The outcomes of this research should also inform future iterations of guidelines on the evaluation of digital scholarship.

Elsewhere, I have observed a dichotomy between the radical discourse of digital humanities — with its frequent talk of revolutions — and its

---

61 Ibid., 57–58.
62 Ibid., 58.
63 See footnote 58.
64 See, for example, Alan Galey and Stan Ruecker, 'How a Prototype Argues', *Literary and Linguistic Computing*, 25.4 (2010), 405–24, https://doi.org/10.1093/llc/fqq021

apparent conformity with the established norms of the academy:[65] for example, the use of (sometimes) blind, pre-publication peer review to evaluate the scholarship it submits to its major journals. One wonders why more experimental and radical approaches to the evaluation of digital scholarship are not being more extensively explored.[66] Is it because of the considerable barriers to open peer review that still exist?[67] Or is it because the price of the field's institutionalisation into the academy has been the abandonment of its radical agenda (if not discourse)? As intimated by Goldfield, peer review is intimately connected with disciplinary identity.[68] Our approaches to the evaluation of digital scholarship in the coming years are of crucial importance, not only in terms of the field's continuing institutionalisation but also in terms of what peer review can reveal about the digital humanities' evolving disciplinary identity.

Bibliography

American Council of Learned Societies, *Our Cultural Commonwealth: The Report of the American Council of Learned Societies Commission on Cyberinfrastructure for the Humanities and Social Sciences* (New York: American Council of Learned Societies, 2006), https://www.acls.org/uploadedFiles/Publications/Programs/Our_Cultural_Commonwealth.pdf

Anderson, Steve, and Tara McPherson, 'Engaging Digital Scholarship: Thoughts on Evaluating Multimedia Scholarship', *Profession* (2011), 136–51 https://doi.org/10.1632/prof.2011.2011.1.136

Brown, Laura, Rebecca Griffiths, and Matthew Rascoff, 'University Publishing in a Digital Age', *The Journal of Electronic Publishing*, 10.3 (2007), https://quod.lib.umich.edu/j/jep/3336451.0010.301?view=text;rgn=main, https://doi.org/10.3998/3336451.0010.301

Chubin, Daryl E., and Edward J. Hackett, *Peerless Science: Peer Review and U.S. Science Policy* (Albany, NY: SUNY Press, 1990).

Finkel, Raphael, et al., 'The Suda On Line (Www.Stoa.Org/Sol/)', *Syllecta Classica*, 11 (2000), 178–90, https://doi.org/10.1353/syl.2000.0005

---

65 See Nyhan and Flinn, *Computation*.
66 See Kathleen Fitzpatrick, 'Revising Peer Review', *Contexts*, 11.4 (2012), 80, https://doi.org/10.1177/1536504212466347
67 See Andy Tattersall, 'For What It's Worth: The Open Peer Review Landscape', *Online Information Review*, 39.5 (2015), 649–63, https://doi.org/10.1108/OIR-06-2015-0182
68 Goldfield, 'An Argument for Single-Author', 372.

Fitzpatrick, Kathleen, 'Peer Review, Judgment, and Reading', *Profession* (2011), 196–201, https://doi.org/10.1632/prof.2011.2011.1.196

—— 'Revising Peer Review', *Contexts*, 11 (2012), 80, https://doi.org/10.1177/1536504212466347

Galey, Alan, and Stan Ruecker, 'How a Prototype Argues', *Literary and Linguistic Computing*, 25.4 (2010), 405–24, https://doi.org/10.1093/llc/fqq021

Goldfield, Joel D., 'An Argument for Single-Author and Similar Studies Using Quantitative Methods: Is There Safety in Numbers?', *Computers and the Humanities*, 27.5–6 (1993), 365–74, https://doi.org/10.1007/BF01829387

'IBM Grants', *ACH Newsletter*, 9.3 (1987).

Jockers, Matthew L., *Macroanalysis: Digital Methods and Literary History*, 1st ed. (Urbana: University of Illinois Press, 2013), https://doi.org/10.5406/illinois/9780252037528.001.0001

Kirschenbaum, Matthew G., 'What Is Digital Humanities and What's It Doing in English Departments?', *ADE Bulletin* (2010), 55–61, https://doi.org/10.1632/ade.150.55

McCarty, Willard, ed., *Humanist Discussion Group Archive (1987–2018)*, http://dhhumanist.org/

—— 'Humanist So Far: A Review of the First Two Months', *ACH Newsletter*, 9.3 (1987).

—— 'HUMANIST: Lessons from a Global Electronic Seminar', *Computers and the Humanities*, 26.3 (1992), 205–22, https://doi.org/10.1007/bf00058618

McGann, Jerome, 'On Creating a Usable Future', *Profession* (2011), 182–95, https://doi.org/10.1632/prof.2011.2011.1.182

McPherson, Tara, 'Scaling Vectors: Thoughts on the Future of Scholarly Communication', *Journal of Electronic Publishing*, 13.2 (2010), https://doi.org/10.3998/3336451.0013.208

Modern Language Association of America, 'Guidelines for Evaluating Work in Digital Humanities and Digital Media', *Modern Language Association* (2012), https://www.mla.org/About-Us/Governance/Committees/Committee-Listings/Professional-Issues/Committee-on-Information-Technology/Guidelines-for-Evaluating-Work-in-Digital-Humanities-and-Digital-Media

Modern Languages Association of America Task Force for Evaluating Scholarship for Tenure and Promotion, *Report of the MLA Task Force on Evaluating Scholarship for Tenure and Promotion* (New York: MLA, 2006), http://www.mla.org/pdf/taskforcereport0608.pdf

Morton, A. Q., 'A Computer Challenges the Church', *The Observer (1901–2003)* (3 November 1963), p. 21.

Moxham, Noah, and Aileen Fyfe, 'The Royal Society and the Prehistory of Peer Review, 1665–1965', *The Historical Journal 61.4* (2018), 863–89, https://doi.org/10.1017/S0018246X17000334

Nowviskie, Bethany, 'Where Credit Is Due: Preconditions for the Evaluation of Collaborative Digital Scholarship', *Profession* (2011), 169–81, https://doi.org/10.1632/prof.2011.2011.1.169

Nyhan, Julianne, and Andrew Flinn, *Computation and the Humanities: Towards an Oral History of Digital Humanities*, 1st ed. (Cham, Switzerland: Springer, 2016).

Olsen, Mark, 'Critical Theory and Textual Computing: Comments and Suggestions', *Computers and the Humanities*, 27.5–6 (1993), 395–400, https://doi.org/10.1007/BF01829390

Pfannenschmidt, Sarah L., and Tanya E. Clement, 'Evaluating Digital Scholarship: Suggestions and Strategies for the Text Encoding Initiative', *Journal of the Text Encoding Initiative* (2014), 7, https://doi.org/10.4000/jtei.949

Raben, Joseph, 'Humanities Computing 25 Years Later', *Computers and the Humanities*, 25.6 (1991), 341–50, https://doi.org/10.1007/bf00141184

Rockwell, Geoffrey, 'On the Evaluation of Digital Media as Scholarship', *Profession* (2011), 152–68, https://doi.org/10.1632/prof.2011.2011.1.152

Schreibman, Susan, Laura Mandell, and Stephen Olsen, 'Introduction', *Profession* (2011), 123–201, https://doi.org/10.1632/prof.2011.2011.1.123

Sinclair, Stéfan, et al., 'Peer Review of Humanities Computing Software', in *ALLC/ACH 2003 — Conference Abstracts* ([n.p.], 2003), pp. 143–45.

Tattersall, Andy, 'For What It's Worth: The Open Peer Review Landscape', *Online Information Review*, 39.5 (2015), 649–63, https://doi.org/10.1108/OIR-06-2015-0182

Unsworth, John, 'Digital Humanities Beyond Representation' (Orlando, FL: University of Central Florida, 2006), http://www.people.virginia.edu/~jmu2m/UCF/

8. Critical Mass

The Listserv and the Early Online Community as a Case Study in the Unanticipated Consequences of Innovation in Scholarly Communication

Daniel Paul O'Donnell

Scholarly communication today exists in a state we might best describe as 'revolutionary stasis'.

On the one hand, it is hard not to be impressed by the disruptive potential that networked computing brings to the way researchers disseminate their results. The development of the Web thirty years ago ushered in a period in which scholars could organise, collaborate, and publish in fundamentally different ways than any time previously. There are new economic models for scholarly publishing, new models for career evaluation and progress, and new understandings of the relationship between scholars and the general public.

However, at the same time, given this revolutionary potential, it is also hard not to be impressed equally by the difficulty these new ideas have had in actually disrupting pre-web ways of working. Indeed, in many cases, traditional markers of success and prestige have become, if anything, even *more* tenacious and entrenched than they were before. Tim Berners-Lee first rolled out the World Wide Web in early 1991.[1] While there has been a slow-but-steady rise in

---

1   Tim Berners-Lee, 'The Original Proposal of the WWW, HTMLized' (1990), http://www.w3.org/History/1989/proposal.html. For a discussion of this conservatism, see Samuel Moore et al., '"Excellence R Us": University Research and the Fetishisation

the number of open access journals, academic publishing is still dominated by the same few presses (e.g. in the humanities: Oxford University Press, Cambridge University Press, Blackwell's, and Routledge).[2] Groups like the Modern Language Association (MLA) have worked to develop new forms of evaluation to accommodate new digital and collaborative forms of scholarship,[3] even as measures of impact that focus on secondary and more traditional markers of use or prestige — such as citation count or journal impact[4] — have become increasingly important through national evaluation schemes such as the United Kingdom's Research Excellence Framework (REF), or the Excellence in Research for Australia (ERA).[5]

In this chapter, I argue that this frustrating state of affairs stems from a misunderstanding of how technological change works in scholarly communication. Although it is very tempting to assume that new platforms, methods of working, and economic models will replace

---

of Excellence', *Palgrave Communications*, 3 (2017), https://doi.org/10.1057/palcomms.2016.105

2 Mikael Laakso and Bo-Christer Björk, 'Anatomy of Open Access Publishing: A Study of Longitudinal Development and Internal Structure', *BMC Medicine*, 10.1 (2012), 124, https://doi.org/10.1186/1741-7015-10-124; Chawki Hajjem, Stevan Harnad, and Yves Gingras, 'Ten-Year Cross-Disciplinary Comparison of the Growth of Open Access and How It Increases Research Citation Impact', *IEEE Data Engineering Bulletin*, 28.4 (2005), 39–47, http://web.archive.org/web/20130814145943/http://eprints.soton.ac.uk/262906/1/rev1IEEE.pdf; Ben Mudrak, 'Scholarly Publishing: A Brief History', *AJE Expert Edge*, http://web.archive.org/web/20190801184847/https://webcache.googleusercontent.com/search?q=cache:d_rJ3pMYOyoJ:https://www.aje.com/arc/scholarly-publishing-brief-history/+&cd=4&hl=en&ct=clnk&gl=ca

3 E.g., Modern Language Association of America, 'Guidelines for Evaluating Work in Digital Humanities and Digital Media', *Modern Language Association* (2012), https://www.mla.org/About-Us/Governance/Committees/Committee-Listings/Professional-Issues/Committee-on-Information-Technology/Guidelines-for-Evaluating-Work-in-Digital-Humanities-and-Digital-Media

4 See Hadas Shema, 'What's Wrong with Citation Analysis?', *Scientific American Blog Network* (1 January 2013), https://blogs.scientificamerican.com/information-culture/whats-wrong-with-citation-analysis/

5 Australian Research Council, *ERA 2018 Submission Guidelines* ([n.p.], 2017), https://web.archive.org/web/20190610203355/https://www.arc.gov.au/sites/default/files/media-assets/era_2018_submission_guidelines.pdf. The British REF 2014 Framework used citation counts in only selected disciplines. See REF, *Assessment Framework and Guidance on Submissions* (Bristol: REF UK, 2011), http://www.ref.ac.uk/2014/media/ref/content/pub/assessmentframeworkandguidanceonsubmissions/GOS%20including%20addendum.pdf. This conservatism is true also of academics not subject to such exercises. See Diane Harley et al., 'The Influence of Academic Values on Scholarly Publication and Communication Practices', *The Journal of Electronic Publishing*, 10.2 (2007), https://doi.org/10.3998/3336451.0010.204

their traditional counterparts, the experience of the last thirty years has demonstrated that new developments in this space tend, instead, to be complementary rather than competitive: that is to say that they introduce additional channels of communication or ways of working rather than replace existing ones. In this sense, our frustration with the degree to which technology has *not* changed scholarly communication may be because we are looking for the change in the wrong places; it is by supplementing and building on what came before, rather than, for the most part, replacing previous methods in fundamental ways, that such innovations ultimately change the way we work.

The Listserv as Case Study

This is perhaps nowhere more apparent than in the history of the academic listserv.[6] Today, the listserv is a work-a-day technology with a very clear role in professional academia: it is one of our principal channels for distributing news and advertisements, calls for papers, conference announcements, changes in address, job vacancies, and technical tips. In the early 1990s, however, the listserv was understood by early-adopting humanists (and humanists were among the very first to use this new technology in a formal academic context) as a potentially revolutionary replacement for a variety of formal academic communication channels, such as the college classroom, the scholarly journal, the academic conference, and even the scholarly society. Exploring the history of this now well-understood technology allows us to see how new dissemination models are incorporated alongside existing channels, thus improving our ability to conduct and report on research without necessarily altering our previous practices. This is 'downstream' from the digital humanities, both in the sense that the first uses and theoretical discussions of the listserv came from the same 'humanities computing' specialists who were responsible for establishing our discipline, and in the sense that studying the impact of such technologies and practices is a core interest of our field. The lessons we learn from the introduction of

---

6   "Listserv" can refer to two different things: the generic idea of an academic mailing list (which is still often called "a listserv[er]") or the specific software utility initially used for their creation. In this chapter, LISTSERV (all capitals) is used to refer to the original software and "listserv" (in the case required by the sentence) to the more generic idea of an academic electronic mailing list.

the listserv, moreover, can help us understand how newer technologies and models with apparently equally revolutionary potential (e.g., the preprint server; social web services such as Humanities Commons, Academia.edu, and ResearchGate; data publication; overlay journals; and scholar-published open access journals) might in fact end up affecting our practice.[7]

First, however, we must go back in time — to rediscover the initial excitement felt by such researchers when they first realised the potential of the listserv.

You've got Mail

Although it seems strange to say it today, one of the most exciting periods in the history of what we now call the 'digital humanities' came in the late 1980s and early 1990s, with the introduction of email and, especially, the LISTSERV mailing list distribution utility to early-adopting humanists.[8]

Email today is understood by most as being, at best, a necessary evil.[9] We complain about it at conferences. There are books and articles

---

7   For preprints, Danielle Padula, 'The Role of Preprints in Journal Publishing', *Scholastica* (7 October 2016), https://blog.scholasticahq.com/post/role-of-preprints-in-journal-publishing/. For overlay journals, Charles Day, 'Meet the Overlay Journal', *Physics Today* (18 September 2015), https://doi.org/10.1063/PT.5.010330; Emilie Marcus, 'Let's Talk about Preprint Servers', *Crosstalk* (3 June 2016), http://crosstalk.cell.com/blog/lets-talk-about-preprint-servers; David Crotty, 'When Is a Preprint Server Not a Preprint Server?', *The Scholarly Kitchen* (19 April 2017), https://scholarlykitchen.sspnet.org/2017/04/19/preprint-server-not-preprint-server/.   On scholar-led journals, Heather Morrison, 'Small Scholar-Led Scholarly Journals: Can They Survive and Thrive in an Open Access Future?', *Learned Publishing: Journal of the Association of Learned and Professional Society Publishers*, 29.2 (2016), 83–88 https://doi.org/10.1002/leap.1015; Bo-Christer Björk, Cenyu Shen and Mikael Laakso, 'A Longitudinal Study of Independent Scholar-Published Open Access Journals', *PeerJ*, 4 (2016), e1990, https://doi.org/10.7717/peerj.1990. On repositories, Paolo Mangiafico, 'Should You #DeleteAcademiaEdu? On the Role of Commercial Services in Scholarly Communication', *LSE Impact of Social Sciences* (1 February 2016),   http://blogs.lse.ac.uk/impactofsocialsciences/2016/02/01/should-you-deleteacademiaedu/; Oya Y. Rieger, 'Opening Up Institutional Repositories: Social Construction of Innovation in Scholarly Communication', *The Journal of Electronic Publishing*, 11.3 (2008), https://doi.org/10.3998/3336451.0011.301

8   'History of LISTSERV', *L-Soft*, http://www.lsoft.com/corporate/history-listserv.asp

9   For a rare opposing view see Peggy Duncan, 'I LOVE Email Campaign Kicks Off October 1st', *Suite Minute Blog by Peggy Duncan* (18 September 2010), http://suiteminute.com/tag/email-culture/

on how to manage it.¹⁰ And it is one of the few constants, along with caffeinated drinks, that are mentioned in almost every blog published during the annual Day in the Life of the Digital Humanities (Day of DH) event, when digital humanities researchers contribute to a disciplinewide account of their activities on a single day.¹¹

Today, we tend to manage, rather than celebrate, our mailing lists. Services like Google Inbox (recently cancelled) or SaneBox offer to filter out news items so that they do not distract from other messages.¹² Some governments even regulate such lists, including those run by non-profits and voluntary groups, by requiring explicit consent from participants.¹³

In the late 1980s and early 1990s, however, both email and mailing lists were understood much more positively. Email addresses and traffic were carried on several different primarily or wholly academic and government networks (BITNET, Internet, and Usenet). Before the National Science Foundation (NSF) began to allow public and commercial access to its Internet in 1992,¹⁴ academic email users rarely, if ever, saw posts from non-academics. 'Spam' (i.e. the use of programs similar to LISTSERV to send commercial emails to multiple recipients) was literally unheard of; the first known commercial mass mailing occurred in April 1994, and the term 'spam' for unwanted messages is thought to have been coined a year earlier.¹⁵ As Robin Peek, who began research for her 1997 dissertation in 1991, argues:

> With these changes, the day of the Internet as a cozy and self-contained academic enclave came to an end. Before 1992, the cultural norms of

---

10   Nick Feamster, 'Time Management Tactics for Academics', *How to Do Great Research* (31 August 2013), https://greatresearch.org/2013/08/31/time-management-tactics-for-academics/; Mike Song et al., *The Hamster Revolution: How to Manage Your Email Before It Manages You*, 1ˢᵗ ed. (Oakland, CA: Berrett-Koehler Publishers, 2008).

11   See 'Initiatives', *centerNet*, http://dhcenternet.org/initiatives

12   Google, 'Inbox by Gmail — the Inbox That Works for You' http://web.archive.org/web/*/https://www.google.com/inbox/; 'Get Rid of Unwanted Email', *SaneBox*, https://www.sanebox.com/home

13   E.g., 'Canada's Anti-Spam Legislation', *Canadian Radio-Television and Telecommunications Commission* (2013), http://crtc.gc.ca/eng/internet/anti.htm

14   See Robin Patricia Peek, 'Early Use of Worldwide Electronic Mailing Lists by Social Science and Humanities Scholars in the United States' (unpublished doctoral dissertation, Syracuse University, New York, 1997), pp. 13–14.

15   Wikipedia contributors, 'History of Email Spam', Wikipedia, The Free Encyclopedia, 2 May 2019, https://en.wikipedia.org/w/index.php?title=History_of_email_spam&oldid=895110052

the Internet, and its oversight by [the] NSF, blocked commercial use of the network. Advertising or charging for use of the Internet was unacceptable. Now, several years later, commercial enterprises freely advertise their wares. The number of Internet users entering the Internet through non-academic providers has grown rapidly. The academic community must now share the Internet with many others who have no direct ties to colleges and universities. Thus, the Internet after 1992 was not the same as the Internet of 1991.[16]

The LISTSERV Revolution

To Peek, who was finishing her dissertation at the near the height of the dot-com bubble,[17] the days of the prelapsarian, university-based Internet in which she began may have seemed 'cozy and self-contained'. But to the pioneers who set up the first academic mailing lists in the second half of the 1980s and early 1990s, the opportunities offered by email and LISTSERV seemed far more expansive and revolutionary. Indeed, to a degree unmatched perhaps even by the flush of excitement that later followed the popularisation of the World Wide Web in the mid-to-late 1990s, academic mailing lists were understood by their early adopters as representing a profoundly disruptive challenge to existing forms of scholarly organisation and communication (see, for example, the historian Erwin K. Welsch, who largely ignores the World Wide Web in his 1994 discussion of what he 'consider[s] to be some important sources for the newly wired historian' but offers instead distinct sections on 'listservers' and Usenet groups).[18]

Teresa M. Harrison and Timothy Stephen are typical of this understanding of the power of what they considered to be an 'altogether new form of discourse'.[19] Like many of those writing and commenting on academic mailing lists at this time, they were already aware of some of the negative qualities that would come to characterise

---

16 Peek, 'Early Use', pp. 13–14.
17 See Wikipedia contributors, 'Dot-Com Bubble', *Wikipedia, The Free Encyclopedia*, 10 June 2019, https://en.wikipedia.org/w/index.php?title=Dot-com_Bubble&oldid=901233426
18 Erwin K. Welsch, 'The Wired Historian: Internet Prospects and Problems', *The Centennial Review*, 38.3 (1994), 479–502 (p. 496).
19 Teresa M. Harrison and Timothy Stephen, 'On-Line Disciplines: Computer-Mediated Scholarship in the Humanities and Social Sciences', *Computers and the Humanities*, 26.3 (1992), 181–93 (p. 190), https://doi.org/10.1007/BF00058616

the format: cliquish behaviour, the domination of discussion by a few frequent posters, problems with focus, and, perhaps especially in their discussion, the difficulty of rewarding professional academics for supporting the public good of academic discussion using a system that is not countenanced by contemporary reward systems.[20]

However, while they are alive to the difficulties associated with the nascent medium, Harrison and Stephen retain a remarkably disruptive and expansive understanding of its possibilities and significance:

> One of the crucial advantages of computer-mediated communication lies in its ability to bring into contact individuals who, due to geography and time constraints, would otherwise be unable to interact with one another on a regular basis [...]. The ability to overcome time and space constraints means that it is far easier to bring individuals together on an interpersonal basis. But CMC [Computer Mediated Communication] has been expected to have a far-reaching and positive effect on the scope and quality of scholarly discourse because it allows groups of individuals to exchange information and resources as well as to interact together in 'computer conferences' that address topics of mutual interest.[21]

The Invisible Seminar

This understanding of the mailing list as a 'conference', 'seminar', 'college', 'journal', or other computer-mediated representation of an existing academic form is characteristic of one of the two main approaches to the design of online, email-based communities taken by pioneers in the mid-1980s and early 1990s.[22] In his account of the

---

20 These and similar issues are discussed throughout the otherwise broadly positive discussions in Willard McCarty, 'HUMANIST: Lessons from a Global Electronic Seminar', *Computers and the Humanities*, 26.3 (1992), 205–22, https://doi.org/10.1007/BF00058618; Patrick W. Conner, 'Networking in the Humanities: Lessons from ANSAXNET', *Computers and the Humanities*, 26.3 (1992), 195–204, https://doi.org/10.1007/bf00058617; Welsch, 'The Wired Historian'. See also the interviews with group participants, conducted in 1992, in Peek, 'Early Use'; for an early (1985) discussion of 'flaming' and other aspects of email etiquette, including appropriate behaviour for mass-distribution lists, see Norman Z. Shapiro and Robert H. Anderson, *Toward an Ethics and Etiquette for Electronic Mail* (Santa Monica, CA: RAND, 1985), http://www.rand.org/pubs/reports/R3283.html
21 Harrison and Stephen, 'On-Line Disciplines', 181–82.
22 See the distinction between 'journal' and other types of lists in Peek, 'Early Use', p. 67. A similar conclusion, using the articles by McCarty and Conner discussed in this chapter, is reached in Avi Hyman, 'Twenty Years of ListServ as an Academic

origins of Humanist, for example, Willard McCarty frequently returns to the appropriateness of the academic seminar — 'a kind of long conversation, convened by a single person but conducted by everyone for mutual enlightenment'[23] — as a metaphor for the mailing list and a guide to his decision-making in its earliest days:

> HUMANIST was formed with little knowledge of networks but strong convictions of what a network for humanists should be like. As editor I was convinced that an e-seminar would gain respect and attract thoughtful people only if it were itself to embody what it sought: mindfulness, and love of language, including respect for spelling, grammar, style, and accuracy of expression.[24]

While the list began as an unmoderated exchange (i.e. where messages from individual subscribers were distributed directly, without any editorial intervention), complaints about information overload and a sudden burst of 'junk mail' stemming from a repeated set of administrative error messages led McCarty to adopt the edited-digest model that characterises the list to this day: posts from subscribers are collected into thematic or generically organised digests, each of which is given a volume and issue number.[25]

Similarly formal approaches characterised other mailing lists established at this time, for example, the LINGUIST List (1990),[26] the *Bryn Mawr Classical Review* (1990),[27] *Postmodern Culture* (1990),[28] and the *Public Access Computer Systems Review* (1989).[29] All of these lists shared the same underlying technology (initially, in most cases, the same LISTSERV utility). All showed a similar commitment to modelling themselves on traditional academic activities or genres,

---

Tool', *The Internet and Higher Education*, 6.1 (2003), 17–24, https://doi.org/10.1016/S1096-7516(02)00159-8

23 McCarty, 'HUMANIST: Lessons', 207. For a discussion of the early history of Humanist, see Julianne Nyhan, 'In Search of Identities in the Digital Humanities: The Early History of Humanist', in *Social Media Archaeology and Poetics*, ed. by Judy Molloy (Cambridge, MA: MIT Press, 2016), pp. 227–24, https://doi.org/10.7551/mitpress/9780262034654.003.0014

24 McCarty, 'HUMANIST: Lessons', 209.

25 Ibid., 210–11.

26 'About LINGUIST List', *The Linguist List*, http://linguistlist.org/about.cfm

27 'About BMCR', *Bryn Mawr Classical Review*, http://bmcr.brynmawr.edu/about.html

28 John Unsworth, personal communication, April 9, 2017.

29 Pat Ensor and Thomas Wilson, 'Public-Access Computer Systems Review: Testing the Promise', *The Journal of Electronic Publishing*, 3.1 (1997), https://doi.org/10.3998/3336451.0003.106

including, characteristically, providing strong editorial moderation and adopting print-era finding aids such as volume and issue numbers (an understandable decision when we remember that before the development of WebCrawler in 1994, there were no full-text search engines; before that, only titles and metadata were indexed).[30] Peek, who excluded journal-type mailing lists from the sample used in her dissertation, nevertheless found the model to be quite widespread at the time, to the extent that even lists that did not formally describe themselves as 'journals', or 'conferences', or 'seminars' nevertheless commonly adopted the model.[31]

The Invisible Water-Cooler

The second main approach to online communities during this period was sometimes described, usually dismissively, as the 'water-cooler'.[32] Because this approach to community-building was later adopted by the major commercial social media platforms (e.g., Facebook, Twitter, and Instagram), and because it has become the default format for new academic mailing lists today (e.g., Digital Medievalist, Global Outlook::Digital Humanities), it appears in many ways more familiar to us than the list-as-'seminar' or 'journal'. Indeed, a Google search shows that 'water-cooler' is now a common element in the name of such mailing lists outside academia.

In the early 1990s, however, this format appears to have been much less common, especially when compared to the 'journal' or 'conference' form. Peek, who had hoped to focus primarily on such 'water-cooler' lists across research disciplines, was ultimately forced to reduce the breadth of her sample due to a lack, in many fields, of suitably active

---

30 'History of Search Engines — Chronological List of Internet Search Engines', *WordStream*, http://www.wordstream.com/articles/internet-search-engines-history; Wikipedia contributors, 'Web Search Engine', *Wikipedia, The Free Encyclopedia*, 9 June 2019, https://en.wikipedia.org/w/index.php?title=Web_search_engine&oldid=901129185

31 Peek, 'Early Use', p. 67. The Text Encoding Initiative mailing lists (initially TEI-L and TEI-TECH) are not included in this discussion as their purpose was always primarily technological rather than humanistic. See Robin Cover, 'SGML/XML Discussion Groups and Mailing Lists', *Cover Pages* (OASIS, Organization for the Advancement of Structured Information Standards, 2001), http://xml.coverpages.org/lists.html

32 Peek, 'Early Use', p. 20.

communities of this type.³³ In other words, what is now the dominant format of the technology was, in its early days, far less popular than its 'revolutionary' cousin.

The characteristic feature of the 'water-cooler' type list is that it does not group its messages topically, or organise or moderate them centrally in an analogy to physical-world academic models. Instead, such groups treat the mailing list as a conversational space in which members ask and answer queries, post announcements, and, in the early years at least, engaged in long-form and short-form discussion, debate, and commentary.

Ansax-l,³⁴ a mailing list for Anglo-Saxonists founded in 1986 as one of the first academic discussion lists in any discipline, adopted this social approach self-consciously. As list-founder, Patrick W. Conner described it a few years after its creation:

> The primary consideration in creating an efficient electronic discussion group is not technical, but social. It is not enough to amass the names of a group of individuals who may or may not be interested in the focus of the list and to tell them how to contact one another; what is needed is a core of participants who will have reasons to correspond with one another, who will introduce more people to the list, and who can be counted upon to become dependent on the discussion group they themselves create [...]
>
> A successful list in the humanities, and probably any list, has to be modeled on an analogy to some social group, such as the extended family or lodge or even college fraternity/sorority. Members have to have full access to one another, and to the group as a whole, to achieve this sort of collegiality, especially when contacts may not be repeated.³⁵

Aware that the more heavily curated approach adopted subsequently by lists such as Humanist more closely mirrored the traditional print formats familiar to established members of the discipline, Conner reports that he worked hard at establishing alternate content and rhetorical models for Ansax-l:

---

33   Ibid., pp. 70–71.
34   Ansax-l is one of two names for this mailing list (the other is Ansaxnet or, in the nomenclature of the time, ANSAXNET). In this paper, I use "Ansax-l" to refer to the actual mailing list (that is to say emails distributed to its subscribers) and "Ansaxnet" to refer to the concept of a mailing list for Anglo-Saxonists.
35   Conner, 'Networking in the Humanities', 196.

> The means [...] of guaranteeing full access socially on ANSAXNET is to de-emphasize titles and honorifics. Of course, I do not introduce nicknames as sometimes happens on public bulletin boards, but I address all members by their first names without the 'dear' and sign myself as 'Patrick' without the 'Sincerely'. For the electronic discussion group to be useful, a graduate student at an American land grant institution, for example, must be capable of exchanging information, without intimidation, with a professor at Oxford or Yale whose work he/she has read.

Perhaps more significantly for the development of the form as a channel for scholarly communication, Conner realised early on that the listserv would also require a different approach to evidence, citation, and relevance:

> For persons used to what I have called the 'print' paradigm, there is no place for ad-lib comments which are not founded on exhaustive bibliographies and thrice-scrutinized logic. Good journals only accept those sorts of scholarly studies, so that many people simply cannot see how an off-the-cuff comment by someone who has not otherwise established his/her credibility in acceptable print media can have value on the list or anywhere else.
> [...I]t is significant that this is an attitude which I have never encountered in new, relatively unpublished scholars. Why should untried scholars not be just as discriminating (or discriminatory) about what they want to read on the list? [...] I think that it is because they are aware that telecommunications exchanges offer a very different model for disseminating professional information, and such an awareness creates a tolerance for chat even when it is irrelevant to their interests. Just as scholars wedded to the print paradigm do not hesitate to read a single work in a book of essays without complaining about the irrelevant items, so the emergent group of telecommunicators have no problem in deleting from their readers materials [that] which the message's subject line shows to be irrelevant.[36]

In keeping with this approach, the models Conner cites for the new list, while occasionally drawing on recognised forms of academic social organisation (e.g., the 'college fraternity/sorority', the 'senior common room', the 'faculty club'), do not include channels or offices traditionally associated with formal scholarly dissemination. In Conner's model for

---

36 Conner, 'Networking in the Humanities', 196, 198.

his list there are no references to examples of journals, newsletters, seminars, or conference sessions, and, above all, there is no official editorial leadership. While he plays the role of moderator and attempts to keep the discussion going, he notes that he tries to avoid putting his stamp on the proceedings:

> ANSAXNET has adopted the model of the modern craft or collector's guild which reasons that the individual's purpose in associating in the first place is the exchange of information. I therefore look for all of the information I can find which might be of interest to an Anglo-Saxonist, sometimes gathering it from other discussion groups and sometimes from print notices which cross my desk, and I post it to the list. If someone sends out a query to which no one else responds, then I respond on-line to make sure that ANSAXNET is perceived by its members as more than a list of addresses which might someday be of use. While on [the] one hand I try to ensure that no one perceives ANSAXNET as merely a personal forum for my ideas, on the other hand I work to maintain an identification with the discussion group, because the perception that a human being is regularly monitoring activities means that all members know that their messages will always be read by at least one other person. I believe that disorder for an electronic discussion group can be defined as the perception by some critical number of members that no one is paying attention. Chat to which no one responds or which is allowed to die helps create this perception, so it is important that observations and queries evoke other observations and queries.[37]

When he does compare his list to a more formal academic channel, namely the academic conference, the reference is to the 'paradiscussion' that takes place in the hallways rather than the papers that take place in the lecture hall.

> The informal transmission of ideas via such rhetoric, which is to be read by everyone but which ostensibly responds to a specific situation or an earlier note, is called 'chat'. Chat is a way to avoid the professional isolation which we often feel at our own universities: it permits interchanges which do not have to begin at the beginning. It serves the social purpose of allowing members who do not know each other personally to establish a kind of 'epistolary' relationship, rather like the 'networking' many persons now say is the primary purpose of attending conferences.[38]

---

37 Ibid., 198–99.
38 Ibid., 197.

What Is It that an Academic Mailing List Disrupts?

I have quoted Conner at length because his emphasis on the importance of para-academic social organisations and practices over more formal scholarly elements in the design of Ansaxnet was as unusual for the time as it is the norm today. When researchers in the 1990s discussed the revolutionary power of 'computer-mediated communication' to transcend time and space, their focus was on the virtual conference panel rather than the virtual coffee break. The disruption they thought was coming (like the disruption we have assumed will accompany subsequent technological developments) involved the disruption of existing formal channels of scholarly dissemination, rather than, as actually turned out to be the case, the informal channels. Where we tend to see the listserv as a cross between a memo and a 'water-cooler', they saw it as a means for disrupting the conference and introducing a new channel for the development and dissemination of research.[39]

Conner too, despite his social models and emphasis on chat, saw real-time research collaboration, the provision of feedback and resources, and debate about specific topics as being necessary for demonstrating the list's relevance to its members. As he noted in one contribution to 'Bicoastal Beowulfians', an early Ansax-L discussion that was itself 'edited, documented, and stored with [...] a journal' in the form of a (print) article published in the *Old English Newsletter*:

> I foresee the day when a topic in which many people participate and offer substance can be subsequently edited, documented, and stored with an electronic journal for larger consumption, as well as being kept on our server for reference. The last thing we want is for folks to think of every word which goes online as a potential article. But some things might grow into that, and now we have the technology to make it relatively painless.[40]

---

39   See among many others, Hyman, 'Twenty Years of ListServ', 20–22; Peek, 'Early Use', pp. 8–9; Harrison and Stephen, 'On-Line Disciplines'; Teresa M. Harrison and Timothy Stephen, *Computer Networking and Scholarly Communication in the Twenty-First-Century University* (Albany, NY: SUNY Press, 1996); McCarty, 'HUMANIST: Lessons'.

40   Contribution by Patrick Conner in Jim Earl et al., 'Bi-Coastal Beowulfians of the '90s: A Curious ANSAXNET Conversation [Excerpted from ANSAXNET, December 1990-February 1991]', *Old English Newsletter*, 24.1 (1990), 36–39 (p. 36), http://www.oenewsletter.org/OEN/archive/OEN24_1.pdf

In actual fact, however, it has been the queries, notices, and other similar types of postings rather than the lengthy opinion pieces or collaborative discussions that have become the core genre of the modern academic list. While the main exception to this, Humanist (a mailing list for researchers in what was at the time known as 'Humanities Computing'), continues both its journal-like paratextual apparatus and its tradition of long and thoughtful posts, a survey of recent postings suggests that the longer contributions come primarily from a small number of contributors (many of whom have been with the list since its beginning) and that the list itself has developed a 'water-cooler'-like flavour as well. Medtext-l (a list for medievalists) had a similar period in which it was characterised by 'long form' posts; this period ended with the passing of its original leader.

Indeed, Conner's sense that 'long-form' exchanges were required to demonstrate the list's relevance never lined up with actual users' interests. In her study of early mailing lists, Peek divides listserv content into four main types of messages: 'Information Exchange', 'Requests for Information', 'Discussion', and 'Technical and Administrative' posts like error messages.[41] Although, as Peek notes, '[p]revious researchers have focussed their efforts on the discussion aspects of computer mediated communication' (i.e. the 'long-form' genres), it was 'Information Exchange' and 'Requests for Information' (the 'short-form' genres that characterise the format today) that, as a rule, provoked the least controversy among subscribers.[42] Both McCarty and Conner similarly indicate that, in practice, it was the 'Discussion'-type posts that provoked more complaints from their membership than anything other than error messages (early mailing list software often had trouble with error and other administrative messages, including subscribers' 'out of office messages' being iteratively reposted to the list).[43] Indeed Ansaxnet became known for its (at the time) particularly lengthy and aggressive

---

41 Peek, 'Early Use', p. 92; Peek treats error messages as being distinct from the other three 'major' categories; her subsequent discussion, as does that of McCarty, however, demonstrates their importance.
42 Peek, 'Early Use', chapter 4.
43 Conner, 'Networking in the Humanities', 197, 199; McCarty, 'HUMANIST: Lessons', 210–12. In the case of Humanist, the 'long form' post, often a digest of multiple replies and responses, became the signature form. It remains unusual among academic mailing lists in this regard and, as noted above, these discussions are, for the most part, prompted by a relatively small group of participants, many of whom have been active in the list leadership since its inception.

discussions — a development that led to the list being temporarily suspended in 1994 or 1995 in a (successful) bid to cool tempers and reset the discussion.[44]

Very few people would imagine today that exchanges on such lists might form the text of a scholarly article, no matter how painlessly it might be put together. The subsequent history of online academic communities confirms the degree to which the 'water-cooler' turned out to be a more productive metaphor than the 'seminar' or 'journal' for these early communities. Academic mailing lists have become a core part of scholarly para- and meta-communication. They are one of the main places where we hear about conferences and calls for papers, arrange conference dinners and meetings, announce publications, and develop community projects. Those who saw the mailing list as a means for adapting existing academic forms such as the journal have, since the advent of the Web, mostly migrated away from email towards web-based publication platforms like Open Journal Systems, which more closely resemble the traditional journal (an exception is LINGUIST List, which retains the trappings of a journal while continuing to use a mailing list for distribution). The lists that remained appear, for the most part, to have given up on this attempt at disruption, focussing instead on filling what turned out to be a previously unmet need for informal communication. Before the listserv, calls for papers were distributed either by advertisements in journals, posters mailed to a network of departments, or by personal (postal) correspondence among friends.[45] With the advent of the listserv, academics organising colloquia or conferences, or putting together special collections or journal issues can use the new technology to reach a far wider network of potential participants in a far shorter period of time, including non-members and people outside their immediate circle of acquaintances. While this was rarely identified by the pioneers of the new technology as a potential benefit, it has turned out, in the end, to represent the real revolutionary development, creating a significant improvement in access for marginalised groups and people working outside the main

---

44 Patrick W. Conner, personal communication, March 28, 2017.
45 See Patrick W. Conner, 'Re: [ANSAX-L] Another Question about Pre-History' (7 April 2017) [electronic mailing list message]; Alison Gulley, 'Re: [ANSAX-L] Another Question about Pre-History' (7 April 2017) [electronic mailing list message].

research centres that in many ways represent a far greater disruption of scholarly practice than the early enthusiasts of the listserv-as-journal hoped to create.[46]

Just as importantly, subsequent generations of online academic communities have picked up where these early lists left off. The various online communities of disciplinary practice that were established in the first decade of the twenty-first century (e.g., Digital Medievalist, 2003; Digital Classicist, 2005; Digital Americanist, 2005) were all built around a 'water-cooler' style mailing list for announcements and requests for information, which was then supplemented by websites/blogs and other, non-email-based, and often offline, academic activities: an online journal (in the case of Digital Medievalist), Wikis, off-line colloquia, conference sessions, and workshops (particularly in the case of Digital Classicist).[47] More recently, a third generation of online academic communities (often loosely associated with previous-generation mailing lists) has been built on commercial social media platforms such as Facebook and Twitter.[48] These involve a similar combination of online networking with off-platform (and often offline) traditional scholarly activity.

Online Communities vs Learned Societies

These second- and especially third-generation communities display no embarrassment about their para-academic, social function. While I am aware of none that uses the metaphor of the 'water-cooler' to

---

46 To argue that the listserv improved access for marginalised groups is not to argue that such groups have achieved equity. There is considerable evidence that people in equity seeking groups still have difficulty gaining access to conference programmes and other channels of research communication. The listserv, however, undoubtedly improved the degree to which calls for papers and other opportunities were distributed to those outside the dominant social networks.

47 Wikipedia contributors, 'Digital Medievalist', *Wikipedia, The Free Encyclopedia*, 29 December 2018, https://en.wikipedia.org/w/index.php?title=Digital_Medievalist&oldid=875874297; Wikipedia contributors, 'Digital Classicist', *Wikipedia, The Free Encyclopedia*, 3 February 2019, https://en.wikipedia.org/w/index.php?title=Digital_Classicist&oldid=881628043; 'About', *Digital Americanists* (2010), http://digitalamericanists.unl.edu/wordpress/about/. For a contemporary discussion, see Gabriel Bodard and Daniel Paul O'Donnell, 'We Are All Together: On Publishing a Digital Classicist Issue of the Digital Medievalist Journal', *Digital Medievalist*, 4 (2008), https://doi.org/10.16995/dm.18, https://journal.digitalmedievalist.org/articles/10.16995/dm.18/

48 E.g., 'Digital Medievalist', *Facebook*, https://www.facebook.com/groups/49320313760/

describe itself, all were more-or-less designed to foster the kind of social communication Conner discusses as his goal in the case of Ansaxnet.

In contrast to the early discussion lists, however, these communities do not, on the whole, see themselves as competing with, replacing, or reconfiguring existing scholarly practices. Harrison and Stephen, for their part, argued that online conferences such as Humanist and Ansaxnet would threaten to replace traditional scholarly societies should they not adopt the same technology and approaches as the (then) new email discussion lists.[49] But while some online communities (particularly the second-generation Communities of Practice such as Digital Medievalist and Digital Classicist, all of which have elected boards) did adopt some of the trappings of the traditional scholarly society, and while some traditional scholarly societies did adopt tools such as the mailing lists used by the newer online communities, the distinction between the two types of communities has remained quite strong. Ansaxnet has not replaced the the International Society for the Study of Early Medieval England (ISSEME) as the main professional body in its discipline any more than Digital Medievalist has replaced the Medieval Academy of America.

As Kathleen Fitzpatrick notes, the traditional societies have their origins in a similar desire to create networking opportunities.[50] But their subsequent development led them to assume primary responsibility for disciplinary certification and credentialing: they became the publishers of the most significant journals, ran the most important conferences, established the prizes that recognised the most important work, hosted the job fairs, and developed disciplinary policies, standards, and formats (including, for example, citation styles) that researchers use in publishing their research, or requesting tenure or promotion. The online communities, on the other hand, have generally avoided all aspects of this certification and policy work. While several organise peer-reviewed conference sessions and colloquia, and, in the case of

---

49 Harrison and Stephen, 'On-Line Disciplines', 190. When Digital Medievalist successfully applied to the Social Sciences and Humanities Research Council's ITST programme for funding in 2005, this was also the main thrust of the very supportive comments by our referees.

50 Kathleen Fitzpatrick, 'Openness, Value, and Scholarly Societies: The Modern Language Association Model', *College & Research Libraries News*, 73.11 (2012), 650–53, https://crln.acrl.org/index.php/crlnews/article/view/8863, https://doi.org/10.5860/crln.73.11.8863.

Digital Medievalist, publish a peer reviewed journal, they have not, for the most part, established annual conferences or prizes, or otherwise engaged in disciplinary standard-setting or gatekeeping. In contrast to the traditional societies, membership in these online communities is invariably free of charge with only a few even accepting donations. Their subscribers tend to hold their online membership alongside, rather than in place of, their membership to the major societies.

In other words, instead of replacing the traditional societies, and with them their domination of the formal channels for the dissemination of scholarly communication, online communities complemented these societies by taking up the networking function they had begun to cede. From the point of view of what they have replaced — the laborious, inefficient, slow, and closed methods of in-group para-disciplinary communication that used to take place by letter, poster, and occasional conference conversations — their impact has been revolutionary. Equally remarkable, however, is the degree to which this is not what the majority of their early proponents predicted they would become. Indeed, in some cases, these proponents actively argued against the possibility that they might become 'no more' than the digital equivalent of a 'water-cooler'.

Same as it Ever Was?
Looking Backwards and Forwards

Technological advances in scholarly communication are almost always initially understood as representing competition, rather than an addition to previously existing techniques, technologies, or economic models.

What the history of the academic mailing list demonstrates for us — or perhaps, more precisely, what the history of disappointed expectations concerning the disruptive potential of academic mailing lists demonstrates — is that such change is far more likely to be complementary than competitive. The mailing list did not replace the academic journal or the scholarly conference, despite the predictions of its early adopters; rather, it created an entirely new, but also entirely complementary, channel for promoting participation in, and distributing information about, such traditional journals and conferences, as well as other more social aspects of academic life. Indeed, as someone who grew up in an academic family but came of academic age myself entirely within the email era, I found it difficult to imagine how the functions currently carried out by the academic listserv and similar social channels were performed before the widespread adoption of email (see above Note 44), and was surprised by the relatively chaotic and ad hoc nature of such face-to-face and postal communications.

It has been thus always, however. As Peek argues, traditional scholarly societies themselves initially developed 'the journal' as a way of improving the efficiency of scientific correspondence:

> For an individual before the seventeenth century the only practical form of communicating over significant distances was the personal letter. In comparison, scholarly journals allowed an individual to communicate more easily and exchange ideas with groups of others.[51]

However, while the journal was initially developed as a way of improving the efficiency of scientific letter writing, it did not, in the end, replace such correspondence: where the letter had originally been about work-in-progress, or exchanging notes or queries as well as final results, by the end of the nineteenth century there had developed a bifurcation, where the journal article became the formal channel for distributing final results while the letter (and later the email) specialised in less-than-final material. Indeed, in this sense, a journal like *Notes and Queries* is an apparent exception that actually proves the rule: despite its title, it is today far more about the publication of (final) notes than (in progress) queries.

The founders of the pioneering academic email lists seem, in turn, to have understood their work as being like the initial journals, that is, an extension of the by-then traditional dissemination solution for final results into a new communication environment. But, in the same way that the journal came to answer a different problem than the scientific correspondence it was supposed to replace, so too the actual impact of the academic mailing list seems, in retrospect, to have been an answer to yet a slightly different question: how do you discover who and what you should pay attention to in an age of effortless dissemination? A filter problem, in other words, rather than a dissemination problem.[52]

---

51 Peek, 'Early Use', p. 6.
52 See O'Reilly. 'Web 2.0 Expo NY: Clay Shirky (shirky.com) It's Not Information Overload. It's Filter Failure', *Youtube*, 19 September 2008, https://www.youtube.com/watch?v=LabqeJEOQyI

This problem, as well as the value of online academic communities as a solution, has only grown more significant as the Web has grown and greater efforts are being made to overcome the academic version of the digital divide.[53] Early accounts of the development of the academic mailing list do not always recognise their true value as simply a means of putting people in touch with each other — a value that has only risen as more and more scholarship is published through non-traditional dissemination channels.

Conclusion

The value of understanding this early history of a technology we now all take for granted is that it may provide a model for understanding some of the frustration we feel with the, at times, surprisingly slow uptake of other 'replacement' technologies, platforms, and models. If the example of the mailing list is anything to go by, we are far more likely to see the long-term survival of traditional means of publication (the book, the subscription journal, the conference, the scholarly society) alongside more novel alternatives (the dynamic book, the overlay journal, or the virtual society) than we are to see any large scale disruption of this space in our lifetimes.

This becomes more interesting, however, when we consider the question of apparently novel forms of dissemination that have been understood to threaten these traditional channels disruptively: the preprint server; social communities and services such as Humanities Commons, Academia.edu, or ResearchGate; data publication; or the scholar-published open access journal.

If the history of the academic listserv is anything to go by, these forms, too, will likely supplement rather than replace the formats we now think of them as competing with. Perhaps what we are seeing here in such new pre-print and offprint distribution mechanisms is the development of formal channels for the non-negotiated distribution of grey literature — methods, data, and results that were previously distributed on a personal basis via email or in person at conferences — in

---

53 See, for example, Daniel Paul O'Donnell, 'In a Rich Man's World: Global DH?', *Dpod Blog* (2 November 2012), http://dpod.kakelbont.ca/2012/11/02/in-a-rich-mans-world-global-dh/

much the same way that conference announcements and calls for papers were before the listserv. At this point, as was true of those looking at the listserv in the early 1990s, it is too early to say precisely how scholarship will change to accommodate these new methods. But as we move downstream, we are more likely to find ourselves in a spreading delta than a churning gorge.

Bibliography

'About', *Digital Americanists* (2010), http://digitalamericanists.unl.edu/wordpress/about/

'About BMCR', *Bryn Mawr Classical Review*, http://bmcr.brynmawr.edu/about.html

'About LINGUIST List', *The Linguist List*, http://linguistlist.org/about.cfm

Australian Research Council, *ERA 2018 Submission Guidelines* ([n.p.], 2017), https://web.archive.org/web/20190610203355/https://www.arc.gov.au/sites/default/files/media-assets/era_2018_submission_guidelines.pdf

Berners-Lee, Tim, 'The Original Proposal of the WWW, HTMLized' (1990), http://www.w3.org/History/1989/proposal.html

Björk, Bo-Christer, Cenyu Shen and Mikael Laakso, 'A Longitudinal Study of Independent Scholar-Published Open Access Journals', *PeerJ*, 4 (2016), e1990, https://doi.org/10.7717/peerj.1990

Bodard, Gabriel, and Daniel Paul O'Donnell, 'We Are All Together: On Publishing a Digital Classicist Issue of the Digital Medievalist Journal', *Digital Medievalist*, 4 (2008), https://doi.org/10.16995/dm.18, https://journal.digitalmedievalist.org/articles/10.16995/dm.18/

'Canada's Anti-Spam Legislation', *Canadian Radio-Television and Telecommunications Commission* (2013), http://crtc.gc.ca/eng/internet/anti.htm

Conner, Patrick W., 'Networking in the Humanities: Lessons from ANSAXNET', *Computers and the Humanities*, 26 (1992), 195–204, https://doi.org/10.1007/bf00058617

—— 'Re: [ANSAX-L] Another Question about Pre-History' (7 April 2017) [Electronic mailing list message].

Cover, Robin, 'SGML/XML Discussion Groups and Mailing Lists', *Cover Pages* (OASIS, Organization for the Advancement of Structured Information Standards, 2001), http://xml.coverpages.org/lists.html

Crotty, David, 'When Is a Preprint Server Not a Preprint Server?', *The Scholarly Kitchen* (19 April 2017), https://scholarlykitchen.sspnet.org/2017/04/19/preprint-server-not-preprint-server/

Day, Charles, 'Meet the Overlay Journal', *Physics Today* (18 September 2015), https://doi.org/10.1063/PT.5.010330

'Digital Medievalist', *Facebook*, https://www.facebook.com/groups/49320313760/

Duncan, Peggy, 'I LOVE Email Campaign Kicks Off October 1st', *Suite Minute Blog by Peggy Duncan* (18 September 2010), http://suiteminute.com/tag/email-culture/

Earl, Jim, et al., 'Bi-Coastal Beowulfians of the '90s: A Curious ANSAXNET Conversation [Excerpted from ANSAXNET, December 1990-February 1991]', *Old English Newsletter*, 24 (1990), 36–39, http://www.oenewsletter.org/OEN/archive/OEN24_1.pdf

Ensor, Pat and Thomas Wilson, 'Public-Access Computer Systems Review: Testing the Promise', *The Journal of Electronic Publishing*, 3.1 (1997), https://doi.org/10.3998/3336451.0003.106

Feamster, Nick, 'Time Management Tactics for Academics', *How to Do Great Research* (31 August 2013), https://greatresearch.org/2013/08/31/time-management-tactics-for-academics/

Fitzpatrick, Kathleen, 'Openness, Value, and Scholarly Societies: The Modern Language Association Model', *College & Research Libraries News*, 73.11 (2012), 650–53, https://crln.acrl.org/index.php/crlnews/article/view/8863, https://doi.org/10.5860/crln.73.11.8863

'Get Rid of Unwanted Email', *SaneBox*, https://www.sanebox.com/home

Google, 'Inbox by Gmail — the Inbox That Works for You', http://web.archive.org/web/*/https://www.google.com/inbox/

Gulley, Alison, 'Re: [ANSAX-L] Another Question about Pre-History' (7 April 2017) [Electronic mailing list message].

Hajjem, Chawki, Stevan Harnad, and Yves Gingras, 'Ten-Year Cross-Disciplinary Comparison of the Growth of Open Access and How It Increases Research Citation Impact', *IEEE Data Engineering Bulletin*, 28.4 (2005), 39–47, http://web.archive.org/web/20130814145943/http://eprints.soton.ac.uk/262906/1/rev1IEEE.pdf

Harley, Diane, et al., 'The Influence of Academic Values on Scholarly Publication and Communication Practices', *The Journal of Electronic Publishing*, 10.2 (2007), https://doi.org/10.3998/3336451.0010.204

Harrison, Teresa M., and Timothy Stephen, *Computer Networking and Scholarly Communication in the Twenty-First-Century University* (Albany, NY: SUNY Press, 1996).

—— 'On-Line Disciplines: Computer-Mediated Scholarship in the Humanities and Social Sciences', *Computers and the Humanities*, 26.3 (1992), 181–93, https://doi.org/10.1007/BF00058616

'History of LISTSERV', *L-Soft*, http://www.lsoft.com/corporate/history-listserv.asp

'History of Search Engines — Chronological List of Internet Search Engines', *WordStream*, http://www.wordstream.com/articles/internet-search-engines-history

Hyman, Avi, 'Twenty Years of ListServ as an Academic Tool', *The Internet and Higher Education*, 6.1 (2003), 17–24, https://doi.org/10.1016/S1096-7516(02)00159-8

'Initiatives', *centerNet*, http://dhcenternet.org/initiatives

Laakso, Mikael, and Bo-Christer Björk, 'Anatomy of Open Access Publishing: A Study of Longitudinal Development and Internal Structure', *BMC Medicine*, 10.1 (2012), 124, https://doi.org/10.1186/1741-7015-10-124

Mangiafico, Paolo, 'Should You #DeleteAcademiaEdu? On the Role of Commercial Services in Scholarly Communication', *LSE Impact of Social Sciences* (1 February 2016), http://blogs.lse.ac.uk/impactofsocialsciences/2016/02/01/should-you-deleteacademiaedu/

Marcus, Emilie, 'Let's Talk about Preprint Servers', *Crosstalk* (3 June 2016), http://crosstalk.cell.com/blog/lets-talk-about-preprint-servers

McCarty, Willard, 'HUMANIST: Lessons from a Global Electronic Seminar', *Computers and the Humanities*, 26.3 (1992), 205–22, https://doi.org/10.1007/BF00058618

Modern Language Association of America, 'Guidelines for Evaluating Work in Digital Humanities and Digital Media', *Modern Language Association* (2012), https://www.mla.org/About-Us/Governance/Committees/Committee-Listings/Professional-Issues/Committee-on-Information-Technology/Guidelines-for-Evaluating-Work-in-Digital-Humanities-and-Digital-Media

Moore, Samuel, et al., '"Excellence R Us": University Research and the Fetishisation of Excellence', *Palgrave Communications*, 3 (2017), https://doi.org/10.1057/palcomms.2016.105

Morrison, Heather, 'Small Scholar-Led Scholarly Journals: Can They Survive and Thrive in an Open Access Future?', *Learned Publishing: Journal of the Association of Learned and Professional Society Publishers*, 29.2 (2016), 83–88, https://doi.org/10.1002/leap.1015

Mudrak, Ben, 'Scholarly Publishing: A Brief History', *AJE Expert Edge*, http://web.archive.org/web/20190801184847/https://webcache.googleusercontent.com/search?q=cache:d_rJ3pMYOyoJ:https://www.aje.com/arc/scholarly-publishing-brief-history/+&cd=4&hl=en&ct=clnk&gl=ca

Nyhan, Julianne, 'In Search of Identities in the Digital Humanities: the Early History of Humanist', in *Social Media Archaeology and Poetics*, ed. by Judy Molloy (Cambridge, MA: MIT Press, 2016), pp. 227–24, https://doi.org/10.7551/mitpress/9780262034654.003.0014

O'Donnell, Daniel Paul, 'In a Rich Man's World: Global DH?', *Dpod Blog* (2 November 2012), http://dpod.kakelbont.ca/2012/11/02/in-a-rich-mans-world-global-dh/

O'Reilly. 'Web 2.0 Expo NY: Clay Shirky (shirky.com) It's Not Information Overload. It's Filter Failure.' *Youtube*, 19 September 2008, https://www.youtube.com/watch?v=LabqeJEOQyI

Padula, Danielle, 'The Role of Preprints in Journal Publishing', *Scholastica* (7 October 2016), https://blog.scholasticahq.com/post/role-of-preprints-in-journal-publishing/

Peek, Robin Patricia, 'Early Use of Worldwide Electronic Mailing Lists by Social Science and Humanities Scholars in the United States' (unpublished doctoral dissertation, Syracuse University, New York, 1997).

REF, *Assessment Framework and Guidance on Submissions* (Bristol: REF UK, 2011), http://www.ref.ac.uk/2014/media/ref/content/pub/assessmentframeworkandguidanceonsubmissions/GOS%20including%20addendum.pdf

Rieger, Oya Y., 'Opening Up Institutional Repositories: Social Construction of Innovation in Scholarly Communication', *The Journal of Electronic Publishing*, 11.3 (2008), https://doi.org/10.3998/3336451.0011.301

Shapiro, Norman Z., and Robert H. Anderson, *Toward an Ethics and Etiquette for Electronic Mail* (Santa Monica, CA: RAND, 1985), http://www.rand.org/pubs/reports/R3283.html

Shema, Hadas, 'What's Wrong with Citation Analysis?', *Scientific American Blog Network* (1 January 2013), https://blogs.scientificamerican.com/information-culture/whats-wrong-with-citation-analysis/

Song, Mike, et al., *The Hamster Revolution: How to Manage Your Email Before It Manages You*, 1st ed. (Oakland, CA: Berrett-Koehler Publishers, 2008).

Welsch, Erwin K., 'The Wired Historian: Internet Prospects and Problems', *The Centennial Review*, 38 (1994), 479–502.

Wikipedia contributors, 'Digital Classicist', *Wikipedia, The Free Encyclopedia*, 3 February 2019, https://en.wikipedia.org/w/index.php?title=Digital_Classicist&oldid=881628043

—— 'Digital Medievalist', *Wikipedia, The Free Encyclopedia*, 29 December 2019, https://en.wikipedia.org/w/index.php?title=Digital_Medievalist&oldid=875874297

—— 'Dot-Com Bubble', *Wikipedia, The Free Encyclopedia*, 10 June 2019, https://en.wikipedia.org/w/index.php?title=Dot-com_bubble&oldid=901233426

—— 'History of Email Spam', *Wikipedia, The Free Encyclopedia*, 2 May 2019, https://en.wikipedia.org/w/index.php?title=History_of_email_spam&oldid=895110052

—— 'Web Search Engine', *Wikipedia, The Free Encyclopedia*, 9 June 2019, https://en.wikipedia.org/w/index.php?title=Web_search_engine&oldid=901129185

9. Springing the Floor for a Different Kind of Dance

Building DARIAH as a Twenty-First-Century Research Infrastructure for the Arts and Humanities

Jennifer Edmond, Frank Fischer, Laurent Romary, and Toma Tasovac

---

Introduction: What's in a Word?

The word *infrastructure* carries the undeniable whiff of heavy engineering, of tar, and gear oil, all accompanied by the sound of a jackhammer. Looking in a dictionary, we will be reminded that infrastructure is basic and foundational, but also that its primary examples are, and remain (in the imagination, if not in reality) in the realm of bricks and mortar: roads, bridges, electricity grids. But the etymology of the word implies nothing of this sort, merely that somewhere below our line of sight, components that support us have been organised. And so, while they may not have the pleasing tangible durability of steel and tarmacadam, marketplaces are equally infrastructural, as are networks of individuals and their knowledge.

Research infrastructures (or RIs) present a particular case where this gap between imagination and function can lead to dissonance. According to one definition, RIs are installations and services that function as 'mediating interfaces' or 'structures "in between" that allow things, people and signs to travel across space by means of more or less standardized paths and protocols for conversion or translation'.[1] A digital research infrastructure is no different: it assembles a mediating set of technologies for research and resource discovery, collaboration, sharing, and dissemination of scientific output.

Infrastructures are not just service providers, however, but also strong cultural and political symbols. From electricity systems in the 1920s, to coal trains in the 1950s, through to the gateways and bridges represented on Euro notes in the present decade, infrastructures have been mobilised repeatedly in broader spheres as symbols and metaphors for the more generalised march of modernisation, integration, and co-operation:[2] engines of change, propelling society into a better and brighter future. Yet, precisely because those 'human-built material links between nations and across borders in Europe [...] predated, accompanied and transcended the "official" processes of political and economic integration',[3] it would be all too tempting — and all too easy — to approach the question of digital research infrastructures uncritically by getting caught up in the moment and embracing the master narratives of efficiency and progress without discussing the larger and more complex implications of institutionalising networked research. A digital infrastructure is not only a tool that needs to be built, it is also a tool that needs to be understood.

Every decade or so, the conceptual framework used by digital humanists to situate the work they do into the landscape of research and its infrastructure is redefined. The idea that the digital could provide quick and easy access to resources drove an early 'access' paradigm. The fact that we could ask new questions about our data drove the rise of a 'methods' paradigm. Now, digital humanities is becoming more mainstream. Furthermore, more of the activities that might be associated

---

1   Alexander Badenoch and Andreas Fickers, 'Introduction Europe Materializing? Toward a Transnational History of European Infrastructures', in *Materializing Europe: Transnational Infrastructures and the Project of Europe*, ed. by Alexander Badenoch and Andreas Fickers (London: Palgrave Macmillan, 2010), pp. 1–23 (p. 2).

2   Badenoch and Fickers, 'Introduction', p. 2; see also Stefan Schmunk et al., 'Interoperabel und partizipativ', in *Digitale Infrastrukturen für die germanistische Forschung*, ed. by Henning Lobin, Roman Schneider, and Andreas Witt (Berlin: De Gruyter, 2018), pp. 53–72, https://doi.org/10.1515/9783110538663-004

3   Badenoch and Fickers, 'Introduction', p. 1.

with traditional as well as digital humanities (such as publishing, where word processing would long have been the 'back end' norm) are becoming overtly digital, and pressures — such as the move toward open science — are bringing technologies for producing and sharing outputs within the consideration of nearly every productive scholar.

In accordance with this, many voices have emerged in the past five years expressing theories about how infrastructure should be understood and delivered for the arts and humanities. In each case, it seems a different role, place, or perspective is offered on what this organised, optimised substrate might offer or should be, whether that is critical cyberinfrastructure,[4] conceptual cyberinfrastructure,[5] tactical infrastructure,[6] or one of any number of emerging characterisations. The rising interest in digital humanities infrastructure might, therefore, be indicative of the long-expected move toward digital humanities becoming an unnecessary compound phrase, as 'digital high-energy physics' would be.

The discussion that follows will take a different approach. This approach entails an examination of practices as much as theories, and an attempt to define infrastructure for the arts and humanities in the digital age — what components it focusses on, what priorities it expresses, how it manifests itself, and how it differentiates itself from its precursors. The discussion will then look specifically at the example of the relatively centralised landscape of research infrastructure in Europe, and the iterative development of the DARIAH ERIC, a consortium of countries committed to a shared programme deployed on behalf of arts and humanities researchers in Europe to build research infrastructure. In particular, the latter half of this chapter will delve into the unique structures and functions this new model of research infrastructure has taken on, taking lessons from the digital humanities, but serving always the disciplines underlying them.

---

4 Alan Liu, 'Toward Critical Infrastructure Studies', Paper Presented at the University of Connecticut, Storrs, 23 February 2017, http://cistudies.org/wp-content/uploads/Toward-Critical-Infrastructure-Studies.pdf

5 Patrik Svensson, 'From Optical Fiber To Conceptual Cyberinfrastructure', *Digital Humanities Quarterly*, 5.1 (2011), http://www.digitalhumanities.org/dhq/vol/5/1/000090/000090.html

6 UC Digital Humanities, 'Dr. Tim Sherratt: Towards a Manifesto for Tactical DH Research Infrastructure', *Youtube*, 2 November 2015, https://www.youtube.com/watch?v=FL5pP2ysjU4

But What *Is* Research Infrastructure?

Trying to extract a succinct definition for *research infrastructure* from existing literature quickly leads to the sense one is listening to the proverbial blind men describing an elephant, each with a different impression of what its purpose might be. In part, this is a result of the many different communities from which these definitions emerge. In order to try and distil a common, consolidated definition, we might start from a set of six published takes on the essence of research infrastructure. Critically, these are derived from six different perspectives: library science, information science, US and EU policy statements, implementation, and cultural theory.[7]

Among these definitions there is very little consensus about what a research infrastructure is comprised of and what its priorities should be. What we can extract from them, however, is a list of components they may have, attributes that may define them, and things they may do. In short, research infrastructures may have the following: facilities, resources, human resources, services, equipment, instruments, collections, archives, databases, structured information systems, grid, computing, software, middleware, information, expertise, standards, policies, tools, knowledge, data, people, a wide user base, and standardised paths and protocols.

At the risk of adding yet another set of elements to the list, we would suggest that this quite varied list can be boiled down to six encompassing categories of assets: tacit and explicit knowledge; networks and communities; software and services; research data collections; labs and instruments; and, finally, buildings and facilities. In and of themselves, none of these assets are inherently infrastructural. However, they can achieve this status by the manner in which they are made available, interoperable, and sustainable. Without these aspects in place, such elements may exist, but within a silo that cannot be shared and reused at a level beyond the walled garden of a project with a limited user group or time limit: a status that renders them unable to meet the minimal requirements of infrastructure.

Returning to our set of definitions, we learn that research infrastructures may be: single-sited, distributed, or virtual;

---

7 For full definitions and citations of the sources used, see Appendix 9.A.

technology-based; shared, unbounded, heterogeneous, open, and evolving; complex agglomerations; diverse; unique; shared broadly; for specific scholarly purposes; sociotechnical systems; an installed base of diverse information technology capabilities; user, operations, and design communities; and more specific than a network, but more general than a tool.

What this multiplicity implies is that research infrastructures are not simply one thing, but exist along a continuum of specialisation, with some able to provide generic support to a wide range of scholars, and others more specialised and serving a smaller group. A possible taxonomy of these levels and types of intervention, offering different assets at different intensities to their user groups, would include technical backbone infrastructures, like GÉANT or national high-speed communications networks for research; standards organisations like the W3C (World Wide Web Consortium), but also the more specific TEI (Text Encoding Initiative) consortium; research centres, which may cover a range of disciplines at a single institution; and, of course, knowledge or memory infrastructures, like museums, libraries, and archives. None of these examples are discipline-specific,[8] however, and one can also observe a model of infrastructure for one or more disciplines that provides bespoke access to a number of assets, and fuses together aspects of these models. There are two other key attributes, however, that any research infrastructures are likely to share: scale and complexity. Without this, a development may be characterised as a tool, useful for a small cohort but unable to intervene widely or in a way that supports community norms without requiring them to adapt significantly to an infrastructurally-enhanced environment.

With a final nod to the existing set of definitions, research infrastructures may undertake to mediate; may allow things, people, and signs to travel across space; may allow individuals to achieve beyond their capacity to know, to do, to see; may support research; and may get 'below the level of the work', a phrase that merits particular attention. The fact that research infrastructures serve research, may seem too obvious to highlight, but many platforms and resources that are hugely useful for the general public or as a teaching resource simply do not have the rigour or richness to support research, and it is difficult

---

8   Except, maybe, for the TEI, whose target audience is essentially humanities scholars.

to retrofit this if it has not built in from the start. The point of inflection, where an infrastructure meets these research needs, is also important, however. It is for this reason that the idea that research infrastructure 'gets below the level of the work' is still worth pausing over almost twenty years after it was first proposed. According to the authors of *Understanding Infrastructure: Dynamics, Tensions and Design*, the ideal state for infrastructure is to be:

> [operating] without specifying exactly how work is to be done or exactly how information is to be processed (Forster and King, 1995). Most systems that attempt to force conformity to a particular conception of a work process (e.g., Lotus Notes) have failed to achieve infrastructural status because they violate this principle (Grudin, 1989; Vandenbosch and Ginzberg, 1996). By contrast, email has become fully infrastructural because it can be used for virtually any work task.[9]

This perspective is not only very much in line with the etymology of the word in question, as discussed in the introduction to this chapter, it also continues to express a key element of how any infrastructure must operate, and the relationship it must have to its users. It also acts as a counterweight to the physicality of the stereotypical images of infrastructures, so common in the imagination and so antithetical to the arts and humanities. As such, it facilitates thought experiments that might define how these two worlds could merge via bridging concepts able to bring to the fore the centrality of knowledge exchange and human interaction in these disciplines. One particular rich field of terminology in this context is that of 'knowledge spaces' or 'knowledgescapes'.

Infrastructures as Knowledge Spaces

According to a pan-European interdisciplinary network of researchers focussed on the potential of the knowledge space as a powerful alternative for knowledge organisation and sharing:

> From libraries to the web; [...] From science maps to interactive knowledge maps; [...] From fundamental research to infrastructures: Physicists, working on complex networks, have developed alternative approaches to knowledge organization by extracting patterns from

---

9   Paul N. Edwards et al., *Understanding Infrastructure: Dynamics, Tensions and Design* (Ann Arbor, MI: Deep Blue, 2007), p. 17, http://hdl.handle.net/2027.42/49353

emerging networks of digitized information. But connections to traditional knowledge orders are rarely discussed, which also hampers their diffusion into information retrieval.[10]

The idea that an infrastructure could facilitate not just the transfer of physical objects or data, but also of knowledge and ideas, is not new. Nonetheless, the idea of the knowledge space opens up a number of intriguing, related semantic spaces. First of all, knowledge spaces are related to the development of 'collective intelligence', a capacity that is 'a much stronger predictor of the team's performance than the ability of individual members',[11] which draws on and increases the 'ability to coordinate tacitly and dynamically'[12] and support 'cognitive or meta-cognitive processes'.[13]

Building infrastructures based upon the fostering of knowledge spaces also gives access to the creation of a 'transactive memory system' (TMS), which can be defined as a 'shared system that individuals in groups develop to collectively encode, share and retrieve information or knowledge in different domains [... for which] there are three behavioural indicators [...]: specialization, credibility and coordination'.[14] This model is therefore highly relevant, as one of the key attributes of infrastructure (as will be discussed in the next section) is scale; and scale requires a division of labour (specialisation), trust between collaborators originating from different epistemic cultures (credibility), and a whole that becomes greater than the sum of its parts (coordination). These capacities of the transactional memory system would enable an infrastructure based on knowledge, even when applied to such a diverse set of disciplines and approaches as the arts and

---

10 Knowescape Project, *Memorandum of Understanding for the Implementation of a European Concerted Research Action Designated as COST Action TD1210: Analyzing the Dynamics of Information and Knowledge Landscapes — KNOWeSCAPE* (Brussels: COST European Cooperation in the Field of Scientific and Technical Research, 2012), p. 5, http://knowescape.org/wp-content/uploads/2013/04/TD1210-e.pdf

11 Anita Williams Woolley, Ishani Aggarwal, and Thomas W. Malone, 'Collective Intelligence in Teams and Organisations', in *Handbook of Collective Intelligence*, ed. by Thomas W. Malone and Michael Bernstein (Cambridge, MA: MIT Press, 2015), pp. 143–57 (p. 143), citing Anita Williams Woolley et al., 'Evidence for a Collective Intelligence Factor in the Performance of Human Groups', *Science*, 330 (2010), 686–88.

12 Williams Woolley, Aggarwal, and Malone, 'Collective Intelligence', p. 147.

13 Ibid., p. 150.

14 Ibid., p. 150.

humanities, to truly facilitate knowledge exchange and the extension of methodologies and fields from 'below the level of the work', as well as to build a peer production-style system of incentives to collaborate, such as the 'intrinsic enjoyment of doing the task, benefits for the contributors from using the software or other innovations themselves, and "social" motivations fed by the presence of other participants on the platform'.[15]

Why Do the Arts and Humanities Need Research Infrastructure?

The technical and material biases that endure in the discourse about research infrastructure also create biases in the general perception of what disciplines require it. However, 'it was in the field of Humanities that the idea of an RI was first born',[16] in the form of the famed Library of Alexandria and its less well-known precursors, of which there is evidence going back thousands of years before the birth of Christ. Even in their digital/social manifestations, the arts and humanities established themselves far earlier than many may believe, with the founding of the TEI Consortium having occurred already in 1987. But the researcher's requirements in the twenty-first century, even in the arts and humanities, are no longer covered completely by the library or archive, even a digital one, and reach far beyond the ambit of a single textual standard (though the TEI is still a major force). Knowledge infrastructures are distinct, and their digital manifestations bring some of their traditional strengths (and weaknesses) to the next generation of their development.

As one of the authors of this chapter has described in more detail elsewhere,[17] the growing accessibility of digital sources has exposed a gap between the infrastructure and its users, which has perhaps always existed, but which is made all the more apparent now because of growing virtual access paradigms and the rise of transnational

---

15 Ibid., p. 157.
16 Claudine Moulin et al., *Research Infrastructures in the Digital Humanities* (Strasbourg: European Science Foundation, 2011), p. 3, http://www.esf.org/fileadmin/user_upload/esf/RI_DigitalHumanities_B42_2011.pdf
17 Jennifer Edmond, 'Tradition and Innovation in the Cendari Research Infrastructure', *Review of the National Center for Digitization 25*, ed. by Zoran Ognjanović (Belgrade: Faculty of Mathematics, University of Belgrade, 2015), pp. 2–9.

approaches to humanities research. The gradual bifurcation between the 'keepers of the sources' and 'facilitators of the activity' was not so much of a problem when access to sources was predicated on occupying the 'space' of a particular holder of rights and knowledge about source material, by which one might mean a library, archive, museum, or indeed a publisher. Cultural heritage institutions are being challenged in their capacity to maintain what is produced by scholars, as production moves from shelves to racks; in their capacity to enable new methodologies in the move beyond reading to 'distant reading'; in their capacity to maintain the high 'up-front' investment required for traditional cataloguing and metadata production; and in their capacity to federate meaningfully across thematic and institutional boundaries.

In short, the challenge of the digital library is to balance old values with the new. In this struggle, we do not want — nor can we afford — to see libraries, museums, and archives forgo their traditional roles as the keepers and protectors of cultural memory. And yet, as the nature of scholarship itself is changing, in the arts and humanities as much as anywhere else, due to the rapid and transformative influence of technology, new, potentially incompatible, requirements for research infrastructure are also emerging. No matter what discipline you work in or how you work, all humanists today must engage with the digital in their work processes, whether their approach engages humanities 'at scale' or in the 'long tail'.

The opportunities are immense, but there are risks as well: 'Faced with the digital "black box", digital models can be imposed upon researchers whose needs in terms of information processing are too often not explained concretely'.[18] The entire field of digital humanities is evolving against the backdrop of global capitalism in its electronic mode, the so-called 'eEmpire', which is sustained by 'a loose assemblage of relations characterized by [...] flexibility, functionality, mobility, programmability, and automation'.[19] It would be naive to think that our fields are immune to the economic and ideological tensions that

---

18   Samuel Szoniecky, 'Ecosystems of Collective Intelligence in the Service of Digital Archives', in *Collective Intelligence and Digital Archives: Toward Knowledge Ecosystems*, ed. by Samuel Szoniecky and Nasreddine Bouhaï (Hoboken, NJ: Wiley, 2017), pp. 1–22 (p. 10), https://doi.org/10.1002/9781119384694.ch1

19   Rita Raley, 'eEmpires', *Cultural Critique*, 57 (2004), 111–50, https://doi.org/10.1353/cul.2004.0014.

characterise information capitalism. It would be even more naive to think we can build expensive, transnational digital research infrastructures that will function in some abstract networked space unburdened by politics and ideology.

Care must be taken, and a community approach adopted. This approach must take into account both the superuser and the marginal case, and must underpin developments as research infrastructure for the arts and humanities seeks to meet the baseline requirements outlined above: scale, openness, durability, and fitness to a broad purpose. It is also important to remember that, in another departure from the old models of the bricks and mortar infrastructure, digital research infrastructure will be a moving target, never able to be viewed as completed or finished. Technology, and its adoption, moves too fast for this to be otherwise. At its best, however, infrastructure will allow any discipline — including, and perhaps particularly, the diverse and atomised arts and humanities — to gain access to networks, data, and knowledge; to achieve greater efficiency and insight in work; to enhance pathways for visibility, reuse and impact; to bring better alignment with shared standards and policy frameworks (such as open science); to increase opportunities for seeking collaborative funding; and to promote long-term sustainability of research outputs.

History of a New Model of RI Development

The rise of a research infrastructure model that could fulfil this significant set of requirements has of course been iterative, but in particular the year 2006 can be pinpointed as being the moment of its consolidation. In this year, two significant publications, one in the US and one in Europe, pointed toward the path along which research infrastructure now develops.

The first of these two publications was the American Council of Learned Societies (ACLS) report on what it called 'humanities cyberinfrastructure', entitled *Our Cultural Commonwealth*, and chaired by John Unsworth.[20] The report was itself a response to an earlier

---

20   American Council of Learned Societies, *Our Cultural Commonwealth: The Report of the American Council of Learned Societies Commission on Cyberinfrastructure for the Humanities and Social Sciences* (New York: American Council of Learned Societies,

one on cyberinfrastructure for science and engineering in the United States, a document generally known as the Atkin's report.[21] While the characteristics of a cyberinfrastructure system for cultural data and investigation described in the ACLS report may have slightly different characteristics from those described elsewhere in this chapter, the eight recommendations given are still remarkably relevant more than a decade later:

1. Invest in cyberinfrastructure for humanities and social science, as a matter of strategic priority.
2. Develop public and institutional policies that foster openness and access.
3. Promote cooperation between public and private sectors.
4. Cultivate leadership in support of cyberinfrastructure from within the humanities and social sciences.
5. Encourage digital scholarship.
6. Establish national centres to support scholarship that contributes to and exploits cyberinfrastructure.
7. Develop and maintain open standards and robust tools.
8. Create extensive and reusable digital collections.[22]

At the same time as Unsworth and his collaborators were developing these recommendations, similar thinking was going on in Europe, albeit not always reaching the same conclusions. In fact, the most prominent representative of what could be seen as a coordinated and comprehensive approach to fulfilling these requirements, namely the Arts and Humanities Data Service (AHDS) in the UK, was defunded in March of 2007. After a decade of supporting the digital aspects of research across the humanities disciplines through its central services

---

2006), https://www.acls.org/uploadedFiles/Publications/Programs/Our_Cultural_Commonwealth.pdf

21 Daniel E. Atkins et al., *Revolutionizing Science and Engineering Through Cyberinfrastructure: Report of the National Science Foundation Blue-Ribbon Advisory Panel on Cyberinfrastructure* (Washington, DC: National Science Foundation, 2003), https://www.nsf.gov/cise/sci/reports/atkins.pdf

22 American Council of Learned Societies, *Our Cultural Commonwealth* (table of contents).

and distributed subject centres, the move left researchers in the UK concerned about the future of support for their work.

While the view from the UK may have seemed opposed to Unsworth and his collaborators' vision, at the European policy level, the future seemed much brighter. A second document published in 2006 was the European Strategic Forum for Research Infrastructure (ESFRI) roadmap,[23] which outlined an initial set of priority investments for pan-European research infrastructures that (it was proposed) member states would build and maintain in a coordinated fashion. On this initial roadmap were three entries with a strong humanities focus: CLARIN,[24] the Common Languages Resources and Technology Infrastructure; EROHS, the European Research Observatory for the Humanities; and DARIAH,[25] the Digital Research Infrastructure for Arts and Humanities.

Of these three, only two ever reached the launch stage: EROHS, like the AHDS, but also like the US-based, Mellon Foundation-funded Project Bamboo, is not currently operational, nor did it ever become so. Of the two remaining humanities research infrastructures on that original roadmap, DARIAH's role and impact is perhaps the more challenging one to understand. CLARIN takes a well-defined community (linguists) and offers them a relatively clear set of tools and services. However, DARIAH serves a more inchoate and diverse community — the arts and humanities writ large — and provides them with something other than a digital library or archive. This task has demanded a different kind of approach, which will be explored below. Nonetheless, DARIAH has been, by every available measure, a successful intervention: after a number of years of preparatory work, it was established as a European Research Infrastructure Consortium (or ERIC) in 2014, and funded from that point on by contributions from the participating member states. In 2016 it was named an ESFRI 'Landmark' project, and its so-called 'operational phase' began in 2019.

Part of DARIAH's success seems to stem from precisely the ways in which it has distinguished itself, even at a structural level, from the other infrastructures on that first ESFRI roadmap. These aspects

---

23 European Strategy Forum for Research Infrastructures, *European Roadmap for Research Infrastructures Report 2006* (Luxembourg: Office for Official Publications of the European Communities, 2006) https://ec.europa.eu/research/infrastructures/pdf/esfri/esfri_roadmap/roadmap_2006/esfri_roadmap_2006_en.pdf
24 *CLARIN ERIC*, https://www.clarin.eu/
25 *DARIAH-EU*, https://www.dariah.eu/

highlight how DARIAH has deployed itself as an infrastructure, but also as a knowledge space for its community. This can be seen, in part, through its relative size at launch: of the first six ERICs launched in 2011–2013, two thirds launched with less than ten national members signed on, a third with only half of that. DARIAH launched with a full fifteen members, and two more joined very shortly after the ERIC had been formed. But critical mass was not the only differentiator. Of those six first infrastructures based on the new European consortial model, only two deployed any sort of in-kind contribution in their funding model, and in those cases the support was specifically earmarked to run national modules or nodes (as in the European Social Survey). In DARIAH, however, the in-kind contributions actually make up a far greater proportion of the member funding requirement than the cash. To be a DARIAH member, countries must organise themselves and their research bases in order to share the tools, data, and knowledge that are developed locally, prioritising reuse and integration over the development of centralised shared services from scratch.

This quirk in the DARIAH funding model reflects the nature of the arts and humanities community and their research, but also the manner in which DARIAH has constructed itself, not merely as, what organisational theorists and economists call a hierarchy, but also as a marketplace.[26] This is a key differentiator given that '[o]ne of the most important ways in which members of groups and organizations coordinate is through their structure. Moreover, the larger the group the more important structure can be in determining the group's effectiveness'.[27] In general, theorists tend to dismiss the marketplace as appropriate to this structuring task, but there are places where it is highly effective: 'If assets are nonspecific, markets enjoy advantages in both production cost and governance cost respects [...] markets can also aggregate uncorrelated demands, thereby realizing risk-pooling benefits; and external procurement avoids many of the hazards to which internal procurement is subject.'[28] If anything can be characterised as a nonspecific asset that meets uncorrelated demands, it is humanities and arts research; and for this, this marketplace model is highly effective. It

---

26 Oliver E. Williamson, 'The Economics of Organization: The Transaction Cost Approach', *American Journal of Sociology*, 87.3 (1981), 548–77.
27 Williams Woolley, Aggarwal, and Malone, 'Collective Intelligence', p. 147.
28 Williamson, 'Economics of Organization', 561.

is also effective when demand is not bilateral, another key aspect of the DARIAH environment.

However, the marketplace aspect of the DARIAH structure is not just a reflection of the privileged place of the in-kind contribution in its funding model. Its entire organisational structure, which is also very different from any other ERIC, reflects this mentality. This is not to say that DARIAH has no hierarchical structure; in fact, it has a very traditional chain of command, with an executive team reporting to a board of directors, who, in turn, answer to a general assembly comprised of representatives of its funders, who each also oversee a national coordinating institution and team. Operating alongside this hierarchy, and feeding into it, however, is a second structure optimised for knowledge sharing and in-flow into the organisation. In this marketplace, a set of four 'Virtual Competency Centres' (VCC) act as gateways and quality assurance nodes for the contributions, not just of the national in-kind contributions (though these have a special status within the information flows), but also from associated research projects and transnational working groups established under the DARIAH umbrella, which will be described in more detail below. The complementarity and links between these two structures can be seen in Figure 1.

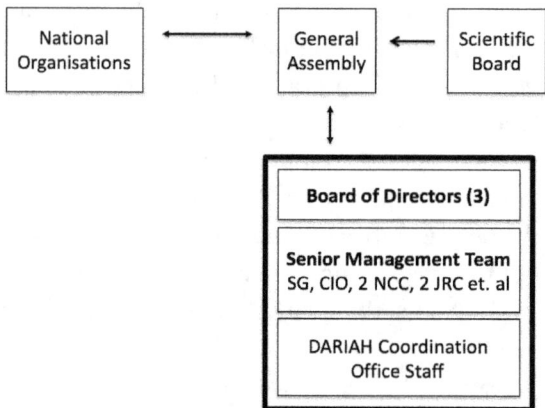

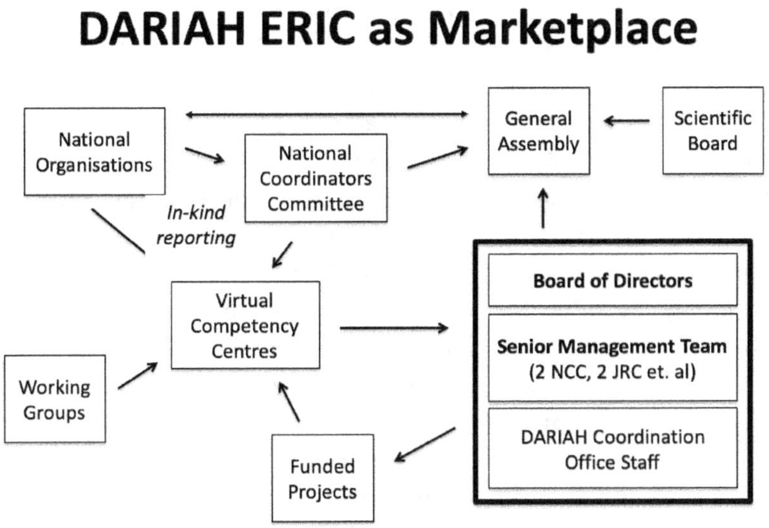

DARIAH ERIC as Marketplace

Fig. 1 DARIAH as Hierarchy and DARIAH as Marketplace. [Figure prepared by the author].

In this way, DARIAH is able to structure its activities so as to meet a quite different and ever-changing set of needs from within its community.

The Activities of the DARIAH ERIC

As outlined above, the DARIAH ERIC serves a broad community, building a new kind of research infrastructure and even sitting between ministries within the European Commission's structure by bringing together elements of the digital agenda, cultural heritage, and education into a research-focussed mission. The need to approach these challenges in a distinct way has been outlined above. However, DARIAH is more than an empty structure. Indeed, the top priority for the national partners, and indeed the researcher-stakeholders, is the impact of DARIAH and the services it delivers.

This is not to say that DARIAH cannot be of benefit merely through the nature of its existence as a body that can speak — if only through the ultimately limited lens of technology — for the needs of the arts and humanities as a whole. Having a mechanism with which to unite the needs of these communities is of benefit in and of itself, creating broadly shared vision and goals within a large community, bringing flexibility and empowerment to local nodes (creating a collective of independent decision makers), harnessing 'collective intelligence', and contributing to the creation of the transactive memory system described above.

DARIAH cannot provide sufficient value to justify the investments made in it merely by existing, and the user-determined worth of infrastructure can make the defining of a clear value proposition difficult in a broad community. For this reason, DARIAH focuses on delivering four flagship initiatives and frames for its activity, allowing it to combine the advantages of top-down and bottom-up development for both the most naive and the most experienced of its users.

These four areas are as follows: a marketplace for validated tools, services, and data aimed at providing inspiration and solutions for the digital aspects of day-to-day research; transnational working groups at the cutting edge of disciplinary and community development; policy and foresight work; and the development of training and career pathways. With these areas, DARIAH seeks to intervene in its environment through a set of 'meta-ideas', which are defined by Paul Romer as:

> [...] ideas about how to support the production and transmission of other ideas. In the seventeenth century, the British invented the modern concept of a patent that protects an invention. North Americans invented the modern research university and the agricultural extension service in the nineteenth century, and peer-reviewed competitive grants for basic research in the twentieth. The challenge now facing all of the industrialized countries is to invent new institutions that encourage a higher level of applied, commercially relevant research and development in the private sector.[29]

Each of these key areas, and the manner in which they can be delivered as an infrastructural service, will be described below.

The DARIAH Marketplace

Of the four key areas on which DARIAH focusses, the most visible is the DARIAH Marketplace for tools, services, data, and knowledge. Structuring DARIAH to function as a marketplace has been an iterative development over the course of the first ten years of the organisation's

---

29   Paul Romer, 'Economic Growth', *Library of Economics and Liberty*, http://www.econlib.org/library/Enc/EconomicGrowth.html

development. What is now called the 'SSH Open Marketplace' will become the showpiece of that set of guiding principles. It addresses a longstanding expression of requirement within the research community that has proven challenging to meet, for an optimal and rich environment for humanists and others to share tools, services, and data. DARIAH's advantage in attempting to meet this need stems from its ability to embed its response in a community framework, harnessing DARIAH's unique in-kind contribution assets, a robust quality control mechanism through its Joint Research Committee and Virtual Competency Centres, reuse cases and contextual material, as well as a reuse imperative, driven by the European policy impulses behind Open Science and in particular the development of the European Open Science Cloud (EOSC).

The development of the EOSC and the manner in which the DARIAH Marketplace frames a bespoke response to it for the arts and humanities provides a good case study for how DARIAH serves its community. The EOSC is being developed as an engine to facilitate the 'most exciting and ground-breaking innovations [that] are happening at the intersection of disciplines'.[30] The vision behind such a grand statement is that by enabling (and encouraging, with the carrot and stick approach) researchers to share not just their completed results in the form of publications, but their research data as well, European researchers will be able to move more fluidly between questions and disciplines, increasing their impact both scientifically and socially. In theory, the EOSC will encompass all disciplines. However, humanists are not always able to share their data, as it may be 'owned' by either an author/creator, or, indeed, a publisher or cultural heritage institution with responsibilities to preserve it, protect it and manage access and use. In addition, humanists do not use data in the same way as other disciplines, and indeed may not even recognise their sources as data. The EOSC will eventually see all researchers who receive European funding required to deposit their research data for reuse, a fact that is of particular concern and interest to DARIAH, as the mismatch between current conceptualisations of data sharing and reuse, including the widely accepted FAIR principles, are largely out of step with existing

---

30 Carlos Moedas, 'The European Open Science Cloud — The New Republic of Letters', *European Commission* (12 June 2017), https://ec.europa.eu/commission/ commissioners/2014-2019/moedas/announcements/eosc-summit-european-open-science-cloud-new-republic-letters_en

humanistic research practices. By building the DARIAH Marketplace as a community-tuned response to EOSC, DARIAH will be able to lead the way, but also mediate between communities of practice currently not in dialogue.

The prospective that the DARIAH Marketplace will be able to manage the risks of epistemic mismatch in a convergent European science system is a strong incentive, but by no means the only one. As DARIAH director Frank Fischer described the vision of the development in a 2017 keynote address,

> Right now, there is no place I could recommend to fellow researchers, where they could go, to look for digital tools or services developed and carved out for the Humanities. Well of course, Google will help you. If you know what you want, that is. But having a central place with tools and services for the Humanities, which is community-driven, where you can find solutions, would be a benefit, and surveys have shown that there's a strong demand for it within the field. A place where you can also count on serendipities, where you can find things you weren't even looking for.
>
> And this is when the DARIAH Marketplace comes into play.
>
> The DARIAH Marketplace is planned as a central, easy-entry place where humanists can find support for the digital aspects of their research. Think of it like a library, but with digital solutions instead of physical books. It will address all humanists, not just those who would regard themselves as digital humanists. It will contain a collection of software, tools, services, datasets, publication repositories and learning & training material and will establish visibility for them.[31]

Through this significant development, DARIAH will deliver on the surprisingly difficult, but long-desired, need for community-based, collective progress in the digital humanities. Delivering fully on this vision will be a worthwhile, albeit decades-long, project: it will significantly contribute to the accessibility of digital approaches in the humanities; it will create visibility, and promote reuse and sustainability for the national contributions DARIAH receives; and it will increase awareness of the barriers to the potential reuse value of digital resources built by researchers for researchers, as well as how to redress these barriers.

---

31 Frank Fischer, 'Towards the DARIAH Marketplace. An Appstore for the Humanities', Keynote Address at DARIAH Innovation Forum, Aarhus, Denmark, November 2–4, 2017.

DARIAH Working Groups

Although DARIAH has always had working groups focussed on particular key task areas, the idea to open up these groups to development from outside the VCC structure only came when the ERIC structure for the membership organisation was launched in 2014. Moving from a top-down to a bottom-up structure has proven highly relevant to community needs however, and the demand for forming these largely unfunded, loose organisations of researchers quickly pushed the number from a handful to over two dozen. Their focus areas are as diverse as they are compelling, from Impactometrix to Federated Identity, and from Women Writers to Natural Language Processing (NLP) approaches.

The working groups provide benefit for both the infrastructure and the participants. They ensure that DARIAH is aware of, as well as meeting, the emergent needs of research communities in the humanities, and, in turn, gives the infrastructure a platform for engagement with them. For the researchers, it provides a non-competitive, non-time-limited, lightweight, transnational mechanism for organising themselves. In addition, there is some funding available to them, and opportunities to meet and showcase their activities are a part of the annual DARIAH meetings, which also encourage exchange and sharing among the groups. More than anything else, however, they facilitate input from the most granular level of the DARIAH stakeholder community: individual researchers and research projects with needs beyond the technical or knowledge landscape they have access to locally.

Policy and Foresight

The average researcher does not care about the process of policy-making, nor should they necessarily have to, as it is a specialist area with its own language, rules, and terms of engagement. That said, the future working conditions researchers will encounter will be determined, at least in part, by policy decisions, and the digital is a particularly pronounced place for research policy to be focussed. For this reason, raising a voice in policy discussions is a key service that DARIAH can provide.

Speaking with a single voice has long been a challenge for the arts and humanities; with so many approaches and disciplines grouped under one term, and not organised in any systematic way, but with a tradition of critical rather than consensual engagement, one can see how hard it would be to forge one. Nonetheless, the concerns of infrastructure provide common ground that makes it easier for the community to agree, or at least to be able to find a common direction to work toward on a number of issues.

The EOSC was mentioned above, and the open science agenda that has given rise to this institution is also a good backdrop against which to consider the kinds of policy engagements DARIAH takes on for the benefit of the arts and humanities research community. A certain amount of this takes the form of actively seeking out and maintaining membership in relevant bodies and projects, such as the Commission's stakeholder body, the Open Science Policy Platform (OSPP);[32] open publishing initiatives such as HIRMEOS,[33] OPERAS,[34] and OpenAIRE;[35] training initiatives like FOSTER+[36] and the OS MOOC;[37] and EOSC-facing initiatives like the EOSC-hub,[38] EOSC Governance Development Forum (EGDF),[39] and the SSH EOSC Cluster, SSHOC.[40] In each of these, DARIAH represents the humanities' perspective, which could be otherwise entirely lost or overlooked, ensuring that the highest-level environment is as friendly as possible to the research communities DARIAH serves, and that those communities are in turn as aware as possible of the trends that will shape their research in the future.

Training, Education, Skills, and Careers

Infrastructures today represent a different model for supporting knowledge creation, but are also, almost as a side effect, developing new models for creating knowledge, differently to their equivalents

---

32   'Open Science Policy Platform', *European Commission*, https://ec.europa.eu/research/openscience/index.cfm?pg=open-science-policy-platform
33   *High Integration of Research Monographs in the European Open Science Infrastructure (HIRMEOS)*, http://www.hirmeos.eu
34   *Open Scholarly Communication in the European Research Area for Social Sciences and Humanities (OPERAS)*, https://operas.hypotheses.org/
35   *OpenAIRE+*, https://www.openaire.eu/
36   *Foster+*, https://www.fosteropenscience.eu/
37   *Open Science MOOC*, https://opensciencemooc.eu/
38   *EOSC-hub*, https://eosc-hub.eu/
39   *EOSCpilot*, https://eoscpilot.eu/
40   *SSHOC*, https://sshopencloud.eu/

in universities and research institutes. They promote different kinds of learning and career development opportunities, often through acculturation processes,[41] but also through certain kinds of overt skills training and formal programmes of access to infrastructures like DARIAH. They are also often a place where careers grow along pathways similar to what has been proposed in the North American conceptualisation of the 'alternate academy'.[42]

Skills acquisition through an infrastructure cannot lend the same formal recognition to participants that one of the many digital humanities doctoral or master's programmes can, but they can serve what may be a more targeted expression of requirement. As Antonijević has described from the results of her ethnography of digital humanists, humanists prefer and learn best in practical settings, when training is embedded in their area of study, and when it develops naturally and interactively.[43]

Into the future, DARIAH expects to see the role of infrastructures continue to rise in importance as a locus for building skills and supporting the new career paths for the research-trained who continue to emerge. Indeed, hierarchies for knowledge creation are in the process of shifting generally (e.g. through the popularisation of 'citizen science'), and applied forms of 'problem-' or 'mission-based' research are on the rise: modes of work that are perhaps uniquely well-supported in and through the new organisational structure for the arts and humanities that infrastructures like DARIAH provide.

Conclusions (and a Few Concerns)

It is undeniable that technology is delivering a sea-change in many aspects of our lives, and arts and humanities research is not immune to this. As a facilitator for this change, the DARIAH research infrastructure

---

41　Geoffrey Rockwell and Stéfan Sinclair, 'Acculturation and the Digital Humanities Community', in *Digital Humanities Pedagogy: Practices, Principles and Politics*, ed. by Brett D. Hirsch (Cambridge: Open Book Publishers, 2012), pp. 177–211, https://doi.org/10.11647/obp.0024.08

42　*#Alt-Academy 01: Alternative Academic Careers for Humanities Scholars*, ed. by Bethany Nowviskie (New York: MediaCommons Press, 2014), http://mediacommons.org/alt-ac/

43　Smiljana Antonijević, *Amongst Digital Humanists: An Ethnographic Study of Digital Knowledge Production* (London, New York: Palgrave Macmillan).

has constituted itself so as to preserve the traditions of the arts and humanities while also encouraging and supporting the uptake of new tools, methods, and opportunities, as well as occupying a unique place in the research landscape. This mission is summarised through the following four points:

- DARIAH serves the arts and humanities research community as an infrastructure, providing a common baseline of access to knowledge and services, but also as an 'interstructure' connecting potentially isolated researchers and fields and creating a fluid basis for the exchange of new insights and methods between them.
- DARIAH complements its stakeholder community, creating a responsive but also protective membrane between the fast-changing world of digital tools and scientific opportunity on the one hand, and the specificity of approaches and contexts that is central to the work of individual humanistic researcher on the other.
- DARIAH's role is far more practical than theoretical. It is comprised of the creators who serve explorers: encouraging and activating, building bridges, drawing up processes, and designing tools that make humanities research more fulfilling and less isolating.
- DARIAH is driven by a passion for the humanities, for their potential to flourish in the digital age, and to serve social, cultural, and economic needs.

The development of the DARIAH ERIC is a case study in harnessing the best of two communities — research infrastructures as originally conceived of in the sciences, and the arts and humanities research base — and merging them in sometimes unexpected ways to create a different, but optimally focussed, proposed range of services. Digital research networks such as DARIAH, however, are also part of a transnational history of materialising Europe, which means that their importance extends beyond strictly scholarly work, opening up a range of central issues, such as:

1. What is the political capital of a digital infrastructure? What is the extent of its sovereignty? And how can we, the community of humanities researchers, make sure that the digital infrastructure — not even the one we are trying to build now, for ours are baby steps, but the future one, the one we hope to see built one day — does not turn from being a power grid into a grid of (hegemonic) power? Sheila Anderson already warned us in 2013 of the uncomfortable alliances research infrastructure development might cause us to make:

   > Although the primary aim of all these infrastructure programmes is to support research, the rhetoric in which they are framed by the funders tends to focus on the economic and political gains to be obtained rather than the advances in knowledge and understanding that they should help to bring about. This emphasis on newness, on innovation, on raising the profile of a country or a continent, conflicts on a number of levels with the reality of infrastructure and its perceived value.[44]

   As the ESFRI roadmap continues to grow, putting pressure on the countries involved in multiple ERICs, and the requirement comes into focus that infrastructures serve industry as well as research, we forget such warnings at our peril, lest we put research at the service of infrastructure rather than vice versa.

2. Infrastructures, in general, have a tendency to disappear out of sight: once the novelty of their implementation wears off, they tend to become invisible or self-evident, taken for granted except for when they fail, inscribed as 'a kind of objective unconscious in our lives'.[45] As we build our digital infrastructures today, we need to prepare for their 'disappearance' tomorrow. We need to think about what type of inherent cultural values and what type of control mechanisms we are programming into digital infrastructures as public institutions before we accept them as an invisible substrate for our work.

---

44 Sheila Anderson, 'What are Research Infrastructures?', *International Journal of Humanities and Arts Computing*, 7.1–2 (2013), 4–23 (p. 7).

45 Dirk van Laak, 'Infra-Strukturgeschichte', *Geschichte und Gesellschaft*, 27.3 (2001), 367–93 (p. 367).

3. The logic of infrastructures is the logic of industrial society: it is based on normativity, mass production, serialisation, and, ultimately, social discipline.[46] As we build a digital infrastructure for the humanities, how do we make sure that we do not end up locking ourselves in, disciplining ourselves to the point that technical protocols become our only destiny, and the limits of our intellectual horizons?

4. When infrastructures remain visible, they usually do so by their absence: in places where they do not exist and where their lack is a very clear indicator of large-scale social inequalities and injustices. We should ask ourselves about the implications of digital infrastructure projects for the dynamics between those who are in and those who are out. Can we create a truly European infrastructure? When will be a good time to start thinking beyond Europe? What are the actual, physical limits of a scientific infrastructure?

DARIAH has come a long way in navigating the dangerous waters of research infrastructure development for the arts and humanities in the digital age. For all the (mistaken, but common) conceptualisations of infrastructure as a one-off capital expenditure, what seems most apparent is that it is a moving target in the digital age, shifting in its ideal focus and service profile as not only the researchers' needs change, but also the environment, the incentives, and the power relations change. As DARIAH moves through its second decade these may be its biggest challenges.

Appendix 9.A: Definitions of Research Infrastructure

[…] facilities, resources or services of a unique nature that have been identified by pan-European research communities to conduct top-level activities in all fields. This definition of Research Infrastructures, including the associated human resources, covers major equipment or sets of instruments, as well as knowledge-containing resources such

---

46 K. J. Beckmann, *Vom Umgang mit dem Alltäglichen. Aufgaben und Probleme der Infrastrukturplanung* (Karlsruhe: Institut für Städtebau und Landesplanung, Universität Karlsruhe, 1988); Mettler-Meibom, B. and C. Bauhardt (Hg.), *Nahe Ferne -fremde Nähe. Infrastrukturen und Alltag* (Berlin: Edition Sigma, 1998).

as collections, archives and databases. Research Infrastructures may be 'single-sited', 'distributed', or 'virtual' (the service being provided electronically). They often require structured information systems related to data management, enabling information and communication. These include technology-based infrastructures such as grid, computing, software and middleware.[47]

Morphologically, digital infrastructures can be defined as shared, unbounded, heterogeneous, open, and evolving sociotechnical systems comprising an installed base of diverse information technology capabilities and their user, operations, and design communities.[48]

In its widest sense, infrastructure allows us, as finite individuals, to achieve beyond our individual capacity to know, to do, to see.[49]

Infrastructure gets

'below the level of the work', i.e. without specifying exactly how work is to be done or exactly how information is to be processed (Forster and King, 1995). Most systems that attempt to force conformity to a particular conception of a work process (e.g., Lotus Notes) have failed to achieve infrastructural status because they violate this principle. By contrast, email has become fully infrastructural because it can be used for virtually any work task.[50]

[…] the term cyberinfrastructure is meant to denote the layer of information, expertise, standards, policies, tools, and services that are shared broadly across communities of inquiry but developed for specific scholarly purposes: cyberinfrastructure is something more specific than the network itself, but it is something more general than a tool or a resource developed for a particular project, a range of projects, or, even more broadly, for a particular discipline. So, for example, digital history collections and the collaborative environments in which to explore and analyze them from multiple disciplinary perspectives might be considered cyberinfrastructure, whereas fiber-optic cables and storage area networks or basic communication protocols would fall below the line for cyberinfrastructure.[51]

---

47 European Strategy Forum for Research Infrastructures, *European Roadmap*, p.16.
48 David Tilson, Kalle Lyytinen, and Carsten Sørensen, 'Research Commentary — Digital Infrastructures: The Missing IS Research Agenda', *Information Systems Research*, 21.4 (2010), 748–59, https://doi.org/10.1287/isre.1100.0318.
49 Jennifer Edmond, 'CENDARI's Grand Challenges: Building, Contextualising and Sustaining a New Knowledge Infrastructure', *International Journal of Humanities and Arts Computing*, 7.1–2 (2013), 58–69 (p. 58), https://doi.org/10.3366/ijhac.2013.0081
50 Paul N. Edwards et al., *Understanding Infrastructure*, p. 17.
51 American Council of Learned Societies, *Our Cultural Commonwealth*, p. 1.

Infrastructures mediate. They are the structures 'in between' that allow things people and signs to travel across space by means of more or less standardised paths and more or less standard protocols for conversation or translation. Thinking of infrastructures as mediating interfaces, that is as points of interaction and translation on material, institutional and discursive levels allows us to get to the heart of the dynamics we seek to capture.[52]

Bibliography

American Council of Learned Societies, *Our Cultural Commonwealth: The Report of the American Council of Learned Societies Commission on Cyberinfrastructure for the Humanities and Social Sciences* (New York: American Council of Learned Societies, 2006) https://www.acls.org/uploadedFiles/Publications/Programs/Our_Cultural_Commonwealth.pdf

Anderson, Sheila, 'What are Research Infrastructures?', *International Journal of Humanities and Arts Computing*, 7.1–2 (2013), 4–23.

Antonijević, Smiljana, *Amongst Digital Humanists: An Ethnographic Study of Digital Knowledge Production* (London: Palgrave Macmillan).

Atkins, Daniel E., et al., *Revolutionizing Science and Engineering Through Cyberinfrastructure: Report of the National Science Foundation Blue-Ribbon Advisory Panel on Cyberinfrastructure* (Washington, DC: National Science Foundation, 2003), https://www.nsf.gov/cise/sci/reports/atkins.pdf

Badenoch, Alexander, and Andreas Fickers, 'Introduction Europe Materializing? Toward a Transnational History of European Infrastructures', in *Materializing Europe: Transnational Infrastructures and the Project of Europe*, ed. by Alexander Badenoch and Andreas Fickers (London: Palgrave Macmillan, 2010), pp. 1–23, https://doi.org/10.1057/9780230292314_1

Beckmann, K. J., *Vom Umgang mit dem Alltäglichen. Aufgaben und Probleme der Infrastrukturplanung* (Karlsruhe: Institut für Städtebau und Landesplanung, Universität Karlsruhe 1988).

CLARIN ERIC, https://www.clarin.eu/

DARIAH-EU, https://www.dariah.eu/

Edmond, Jennifer, 'CENDARI's Grand Challenges: Building, Contextualising and Sustaining a New Knowledge Infrastructure', *International Journal of Humanities and Arts Computing*, 7.1–2 (2013), 58–69, https://doi.org/10.3366/ijhac.2013.0081

—— 'Tradition and Innovation in the Cendari Research Infrastructure', *Review of the National Center for Digitization 25*, ed. by Zoran Ognjanović (Belgrade: Faculty of Mathematics, University of Belgrade, 2015), pp. 2–9.

---

52   Badenoch and Fickers, 'Introduction', p. 2.

Edwards, Paul N., et al., *Understanding Infrastructure: Dynamics, Tensions and Design* (Ann Arbor, MI: Deep Blue), http://hdl.handle.net/2027.42/49353

EOSCpilot, https://eoscpilot.eu/

EOSC-hub, https://eosc-hub.eu/

European Strategy Forum for Research Infrastructures, *European Roadmap for Research Infrastructures Report 2006* (Luxembourg: Office for Official Publications of the European Communities, 2006), https://ec.europa.eu/research/infrastructures/pdf/esfri/esfri_roadmap/roadmap_2006/esfri_roadmap_2006_en.pdf

Fischer, Frank, 'Towards the DARIAH Marketplace. An Appstore for the Humanities', Keynote Address at DARIAH Innovation Forum, Aarhus, Denmark, November 2–4 2017.

Foster+, https://www.fosteropenscience.eu/

High Integration of Research Monographs in the European Open Science Infrastructure (HIRMEOS), http://www.hirmeos.eu

Knowescape Project, *Memorandum of Understanding for the Implementation of a European Concerted Research Action Designated as COST Action TD1210: Analyzing the Dynamics of Information and Knowledge Landscapes — KNOWeSCAPE* (Brussels: COST European Cooperation in the Field of Scientific and Technical Research, 2012), http://knowescape.org/wp-content/uploads/2013/04/TD1210-e.pdf

Liu, Alan, 'Toward Critical Infrastructure Studies', Paper Presented at the University of Connecticut, Storrs, 23 February 2017, http://cistudies.org/wp-content/uploads/Toward-Critical-Infrastructure-Studies.pdf

Mettler-Meibom, B., and C. Bauhard, eds., *Nahe Ferne — fremde Nähe. Infrastrukturen und Alltag* (Berlin: Edition Sigma, 1998).

Moedas, Carlos, 'The European Open Science Cloud — The New Republic of Letters', *European Commission* (12 June 2017), https://ec.europa.eu/commission/commissioners/2014-2019/moedas/announcements/eosc-summit-european-open-science-cloud-new-republic-letters_en

Moulin, Claudine, et al., *Research Infrastructures in the Digital Humanities* (Strasbourg: European Science Foundation, 2011), http://www.esf.org/fileadmin/user_upload/esf/RI_DigitalHumanities_B42_2011.pdf

Nowviskie, Bethany, ed., *#Alt-Academy 01: Alternative Academic Careers for Humanities Scholars* (New York: MediaCommons Press, 2014), http://mediacommons.org/alt-ac/

OpenAIRE+, https://www.openaire.eu/

Open Scholarly Communication in the European Research Area for Social Sciences and Humanities (OPERAS), https://operas.hypotheses.org/

'Open Science Policy Platform', *European Commission*, https://ec.europa.eu/research/openscience/index.cfm?pg=open-science-policy-platform

Open Science MOOC, https://opensciencemooc.eu/

Raley, Rita, 'eEmpires', *Cultural Critique*, 57 (2004), 111–50, https://doi.org/10.1353/cul.2004.0014

Rockwell, Geoffrey, and Stéfan Sinclair, 'Acculturation and the Digital Humanities Community', in *Digital Humanities Pedagogy: Practices, Principles and Politics*, ed. by Brett D. Hirsch (Cambridge: Open Book Publishers, 2012), pp. 177–211, https://doi.org/10.11647/obp.0024.08

Romer, Paul, 'Economic Growth', *Library of Economics and Liberty*, http://www.econlib.org/library/Enc/EconomicGrowth.html

Schmunk, Stefan, et al., 'Interoperabel und partizipativ' in *Digitale Infrastrukturen für die germanistische Forschung*, ed. by Henning Lobin, Roman Schneider, and Andreas Witt (Berlin: De Gruyter, 2018), pp. 53–72, https://doi.org/10.1515/9783110538663-004

SSHOC, https://sshopencloud.eu/

Svensson, Patrik, 'From Optical Fiber to Conceptual Cyberinfrastructure', *Digital Humanities Quarterly*, 5.1 (2011), http://www.digitalhumanities.org/dhq/vol/5/1/000090/000090.html

Szoniecky, Samuel, 'Ecosystems of Collective Intelligence in the Service of Digital Archives', in *Collective Intelligence and Digital Archives: Toward Knowledge Ecosystems*, ed. by Samuel Szoniecky and Nasreddine Bouhaï (Hoboken, NJ: Wiley, 2017), pp. 1–22, https://doi.org/10.1002/9781119384694.ch1

Tilson, David, Kalle Lyytinen, and Carsten Sørensen, 'Research Commentary — Digital Infrastructures: The Missing IS Research Agenda', *Information Systems Research*, 21.4 (2010), 748–59, https://doi.org/10.1287/isre.1100.0318.

UC Digital Humanities, 'Dr. Tim Sherratt: Towards a Manifesto for Tactical DH Research Infrastructure', *Youtube*, 2 November 2015, https://www.youtube.com/watch?v=FL5pP2ysjU4

Van Laak, Dirk, 'Infra-Strukturgeschichte', *Geschichte und Gesellschaft*, 27.3 (2001), 367–93.

Williams Woolley, Anita, Ishani Aggarwal, and Thomas W. Malone, 'Collective Intelligence in Teams and Organisations', in *Handbook of Collective Intelligence*, ed. by Thomas W. Malone and Michael Bernstein (Cambridge, MA: MIT Press, 2015), pp. 143–57.

Williamson, Oliver E., 'The Economics of Organization: The Transaction Cost Approach', *American Journal of Sociology*, 87.3 (1981), 548–77.

10. The Risk of Losing the *Thick Description*

Data Management Challenges Faced by the Arts and Humanities in the Evolving FAIR Data Ecosystem[1]

Erzsébet Tóth-Czifra

Realising the Promises of FAIR within Discipline-Specific Scholarly Practices

Since their inception in 2014, the FAIR principles (findability, accessibility, interoperability, and reusability) have come a long way in serving the global need for generic guidelines for data management and stewardship.[2] Addressing one of the grand challenges of scientific innovation, namely the need for infrastructure that supports the reuse of scholarly data, the FAIR principles have become increasingly influential since their formulation (created by a wide range of stakeholder groups who came together)[3] as a framework for the enhancement and optimisation of the digital ecosystem surrounding scholarly data publication.

---

1   I wish to thank Laurent Romary and Jennifer Edmond for their invaluable suggestions and comments on an earlier version of this manuscript.
2   Mark D. Wilkinson et al., 'The FAIR Guiding Principles for Scientific Data Management and Stewardship', *Scientific Data*, 3 (2016), https://doi.org/10.1038/sdata.2016.18
3   *Jointly Designing a Data FAIRPORT*, Workshop at Lorentz Center@Snellius, Leiden, 13–16 January 2014, https://www.lorentzcenter.nl/lc/web/2014/602/info.php3?wsid=602

The strong need for guidelines to enable and incentivise sustainable, connected, easily accessible, and cost-effective models of scholarly data curation was clearly reflected in the reception of the FAIR principles. The wide embrace and support for FAIR by governments, policy-makers, governing bodies, and funding bodies has not only made FAIR data, or 'FAIRification', a synonym for high-quality scientific data production, but has also fast-tracked the principles so they could make their way into global policies worldwide,[4] despite the many open questions their implementation leaves behind, and the palpable lack of agreed implementation plans and models at the level of different disciplines.

Considering how deeply they are embedded in the landscape of European scientific innovation and policy, the FAIR principles have the potential to make a substantial impact on the future landscape, as well as to shape the underlying dynamics of knowledge creation for the better. This chance, however, can easily be missed if the specific dynamics of scientific production in the humanities are not addressed in their discipline-level implementation.

With the goal of making FAIR meaningful, and helping it to realise its promises in an arts and humanities context, this paper describes some of the defining aspects underlying the domain-specific, epistemic processes that pose challenges to the FAIRification of knowledge creation in arts and humanities. In particular, by applying the FAIR principles to arts and humanities data curation workflows, it is demonstrated that, contrary to the principles' general scope and deliberately domain-independent nature, the principles have been implicitly designed according to underlying assumptions about how knowledge creation operates and communicates. In the following sections three such assumptions are addressed: first, that scholarly data or metadata is digital by nature;[5] second, that scholarly data is always

---

[4] See, for example, European Commission, Directorate-General for Research & Innovation, *H2020 Programme Guidelines on FAIR Data Management in Horizon 2020* (26 July 2016), http://ec.europa.eu/research/participants/data/ref/h2020/grants_manual/hi/oa_pilot/h2020-hi-oa-data-mgt_en.pdf; or Australian FAIR Access Working Group, *Policy Statement on FAIR Access to Australia's Research Outputs*, https://www.fair-access.net.au/fair-statement

[5] See the 'Preamble' of the principles of: FORCE11, 'Guiding Principles for Findable, Accessible, Interoperable and Re-Usable Data Publishing Version B1.0', *FORCE11* (2014), https://www.force11.org/fairprinciples, where the eScience ecosystem is clearly indicated as being the domain of FAIR data management.

created and, therefore, owned by researchers;[6] and third, that there is wide community-level agreement on what can be considered to be scholarly data. The problems surrounding such assumptions in arts and humanities are the cornerstones for reconciling disciplinary traditions with FAIR data management. By addressing these assumptions one by one, this chapter contributes to a better understanding of the discipline-specific needs and challenges in data production, discovery, and reuse. These considerations may facilitate the inclusive and optimal implementation of high-level principles in a way that will serve to make the arts and humanities' disciplines flourish, rather than imposing limitations on their epistemic practices.

A Cultural Knowledge Iceberg, Submerged in an Analogue World

There is a fundamental difference between the epistemic cultures of STEM (science, technology, engineering, and mathematics) and those of the arts and humanities: namely, that in the arts and humanities the wide range of scholarly information artefacts, works of art, written documents of all sorts, recordings, annotations etc. — all of which can be broadly referred to as research data (in the sense used by Margaret E. Henderson)[7] — are not the autonomous products of research projects, but rather are deeply embedded in the cultural memory of Europe as well as the cultural and social practices of the institutions that preserve, curate, and (co)produce them. These institutions, commonly referred to as cultural heritage or GLAM (galleries, libraries, archives, museums) institutions — ranging from national libraries and archives down to small village museums or administrations — are typically not part of

---

6   Note that in the 'Preamble' there is no reference to data providers and data curators other than researchers (such as private or publicly funded providers of medical data, or curators of cultural heritage) nor are they mentioned among the stakeholders.

7   *Data Management: A Practical Guide for Librarians* (Lanham, MD: Rowman & Littlefield, 2016): 'Research data is data that is collected, observed, or created, for purposes of analysis to produce original research results' (p. 2). Other data definitions in a humanities context are more restrictive, for example, that of Christof Schöch (2013) in Christof Schöch, 'Big? Smart? Clean? Messy? Data in the Humanities', *Journal of Digital Humanities*, 2.3 (2013), http://journalofdigitalhumanities.org/2-3/big-smart-clean-messy-data-in-the-humanities/. As we will note later in this paper, the notion of research data is far from being straightforward in the arts and humanities.

the institutional landscape of academia. Despite this, the digital research ecosystem poses many challenges connected to the exploration and exploitation of the material and collections they hold; we do not need to get very far into the FAIR acronym to recognise these challenges.

The fact that these cultural sources and their enrichments are not merely representations of history, but also come with their own histories in terms of their creation and provenance, has serious implications regarding their visibility and shareability. Most importantly, the long tradition of cultural heritage data curation determines the way in which cultural resources are made available. According to a Europeana Foundation white paper from 2015, only ten percent of European cultural heritage is digitally available (300 million objects).[8] Therefore, the vast majority of cultural heritage data remain invisible on the digital horizon, which serves as the default domain of FAIR and scientific data management. Despite the combined digitisation efforts in Europe,[9] these numbers suggest that, for the foreseeable future, arts and humanities research will retain its hybrid nature, and encompass varying degrees of digital and analogue elements, thus calling for both automated and manual workflows and practices.

To give an example illustrating how much effort and investment is required to satisfy the basic requirements of data being digital in a cultural heritage context, Samuelle Carlson and Ben Anderson refer to two digitisation projects as cases in point: the CurationProject, which aimed at digitising and making available for study the records of a collection of more than 750,000 artefacts and 100,000 field photographs that had been collected since 1884; and the AnthroProject, where anthropological materials (including fieldwork notes, images, maps,

---

8 *Transforming the World with Culture: Next Steps on Increasing the Use of Digital Cultural Heritage in Research, Education, Tourism and the Creative Industries*, ed. by Beth Daley (The Hague: Europeana Foundation, 2015), https://pro.europeana.eu/files/Europeana_Professional/Publications/Europeana%20Presidencies%20White%20Paper.pdf. See also the same numbers in Fig. 3.6 in Gerhard Jan Nauta and Wietske van den Heuvel, *Survey Report on Digitisation in European Cultural Heritage Institutions 2015* (The Hague: DEN Foundation/Europeana/ENUMERATE, 2015), http://enumeratedataplatform.digibis.com/reports/survey-report-on-digitisation-in-european-cultural-heritage-institutions-2015/detail

9 European Commission, *Digitisation, Online Accessibility and Digital Preservation. Report on the Implementation of Commission Recommendation 2011/711/EU (2013–2015)*, http://ec.europa.eu/information_society/newsroom/image/document/2016-43/2013-2015_progress_report_18528.pdf

and texts) from a range of countries were digitised and distributed through an online database and via DVDs.¹⁰ In both projects, the major challenge was to build a well-structured, searchable database from their rather heterogeneous sources and records. This aim was realised as a rather long-term goal for both projects: the progressive digitisation, curation, and systematic documentation took thirty years in both cases.

Taking a step further towards findability, although digitisation is a preliminary first step in sharing knowledge, it alone does not guarantee the visibility and accessibility of cultural heritage data outside the walls of their hosting institutions. The aforementioned Europeana survey reveals that only one third (thirty-four percent) of digitised cultural heritage resources are currently available online, with barely three percent of these works suitable for real creative reuse; meaning, only this three percent has the chance to fulfil the discipline-specific measures of being FAIR.¹¹

There are a number of cultural, social, legal, technical, and economic reasons that explain this small percentage of truly reusable cultural heritage data. These circumstances impact greatly on the working conditions of not only librarians, museologists, and archivists but also that of scholars who want to reuse and share data and content relevant to their research.

Legal Problems that Are Not Solely Legal Problems

The biggest obstacle in the productive reuse of digitised cultural heritage resources — from which many others derive — is the legal and ethical restrictions in which the usage conditions of cultural heritage sources are embedded. Determining the ownership status of research that is based on such material poses challenges in many cases. This is because the ownership status of research is, on some level, shared between the researcher who carries out the scientific analysis on the source materials, the institution that hosts and curates this material, and the people and cultures who give rise to the objects in question (e.g., photographers, and also the subjects of the photographs). Establishing

---

10 Samuelle Carlson and Ben Anderson, 'What *Are* Data? The Many Kinds of Data and their Implications for Data Re-Use', *Journal of Computer-Mediated Communication*, 12.2 (2007), 635–51, https://doi.org/10.1111/j.1083-6101.2007.00342.x

11 Daley, ed., *Transforming the World with Culture*, p. 9.

precise conditions for reuse on the basis of such a complex web of claims is, therefore, not an easy task.¹²

In addition to this complexity, provenance trails (i.e. a documented ownership and curation history of an artefact) are often embedded in historical practices, in particular in eras or contexts when the legal-ethical framework that defines present-day data exchange was either non-existent or irrelevant. Obviously, those handling these data could not know in advance that some information — for example, attribution or consent from the rights holders — needed to be collected: this requirement was only brought about by the digital age. Tracing back the provenance of such records is a time-consuming and difficult process filled with uncertainties and lack of clarity, especially in the case of collections inherited from other institutions.¹³

Furthermore, even in cases where the entity holding the legal right is clearly identifiable, given the great deal of legal uncertainty and variety present at the intersection of differing national legislations, and the changing landscape of intellectual property rights (IPR), in many cases researchers and curators are having difficulty 'translating' the legal statuses and license information of materials into research and publication workflows and terms of use. For instance, the legal statement 'In copyright, non-commercial use only' raises the question of where commercial use begins. Visual material under this legal status can certainly be integrated into PhD dissertations, but what about republishing such material on the researcher's website or in scholarly monographs?

The broad investigations of archival practices conducted within the framework of the Knowledge Complexity (KPLEX) project by Mike Priddy

---

12   To illustrate this complexity, let us cite here two examples from Carlson and Anderson's two aforementioned case studies: '[A researcher] has put a picture on the cover of a publication. He could be fined for that [by the community it originated from], because the artifact [sic] shows a ritual/secret process.'; and 'during her fieldwork in Malaysia, there was a photo collection (of a former local museum) that they wanted to sell to us. There were photos by tourists, army officers, etc. They think that they own every photo, but in our sense the photographer owns it, and we can therefore not show it' ('What *Are* Data?', 643).

13   This legal uncertainty in the identification of the legal statuses of cultural heritage material is clearly represented in the fact that in the Rights Statements framework, which has been designed specifically for cultural heritage data where the rights holder and the data provider are not always the same entities, four of the twelve standardised rights statements refer to unclear legal statuses. These are: 'In Copyright/Rights-holder(s) Unlocatable or Unidentifiable', 'Copyright Not Evaluated', 'Copyright Undetermined', and 'No Known Copyright'. See *Rights Statements for in Copyright Objects*, http://rightsstatements.org/en/

and Nicola Horsley reveal how such legal restrictions also affect technical and cultural aspects of data sharing in the cultural heritage domain.[14] In the context of developing support for interoperability frameworks via metadata standards and computational research methods, it is important to recognise that perceived or substantive legal barriers not only impact on the barriers for the reuse of content, but may prevent institutions from online metadata sharing as well. The identity of individuals or groups are often so deeply inscribed in the data that not even the highest level of abstraction can shield them. For example, some collection descriptions cannot be made available online because they contain biographical information about the person who donated them.

As the following excerpt from one of the interviews conducted in the KPLEX project indicates, such difficulties are either slowing down the standardisation procedure, increasing the manual curation effort required to produce sufficient and safe metadata, or simply preventing metadata sharing. This is especially problematic in the context of the FAIR recommendation that metadata should be open by default, even in cases of sensitive data.[15]

> [T]hese kinds of problems asked us to be able to make a choice between the collections, the metadata, which can be shared and the other ones and that took a lot of time. We weren't able to do that automatically, so these kinds of things, and it was totally impossible for us. So, for example, for [portal], to share metadata or to share documents with [portal]. It wasn't possible because of copyright issues or privacy issues.[16]

The need to fulfil legal requirements and to avoid the risk of penalties drives a conservative stance where there may be any uncertainty or grey area, and incentivises the practices of reduced sharing or holding data back out of a fear of lawsuits against, and legal liability of, the

---

14   Mike Priddy and Nicola Horsley, 'Deliverable D3.1 Report on Historical Data as Sources', *KPLEX* (2018), https://kplexproject.files.wordpress.com/2018/06/kplex_deliverable-d3-1.pdf. KPLEX is a Horizon 2020 project aimed at investigating ways in which a focus on 'big data' in ICT research elides important issues about the information environment we live in. The project focuses on four main themes: toward a new conceptualisation of data; hidden data and the historical record; data, knowledge organisation and epistemics; and culture and representations of system limitations.

15   Simon Hodson et al., 'Turning FAIR Data into Reality: Interim Report from the European Commission Expert Group on FAIR Data', *Zenodo* (2018), https://doi.org/10.5281/zenodo.1285272: 'The basic core is proposed as discovery metadata, persistent identifiers, and access to the data, or, at a minimum, metadata' (p. 57).

16   Priddy and Horsley, 'Deliverable D3.1 Report', p. 65.

respective institutions. The lack of a clear definition regarding the legal barriers puts a large portion of cultural heritage material into a minefield that neither practitioners in cultural heritage institutions nor scholars are willing to step into. The abandonment of certain research questions due to legal uncertainty, and the lack of accurate, transparent, and easily understandable conditions of access to documents, is an even bigger obstacle to FAIRification in the cultural heritage domain than the institution of legal protection that it aims to serve.

Case Study: The Removal of Photos from the CENDARI Project's Archival Research Guides due to a Lack of Information on their Reuse Conditions

The following case study from the CENDARI[17] project illustrates how legal, cultural, and data-management dimensions of non-transparency can lock away valuable and relevant cultural data so they cannot be reused, shared, and therefore sustainably preserved in the collective practices of heritage maintenance.

In February 2016, at the time of finalising the publication of CENDARI's Archival Research Guides,[18] scholars working on First World War materials were faced with a situation in which the ownership status of the illustrative images (found on the internet) was so unclear and inaccessible (even after detailed and repeated checks) that eventually the images in question had to be left out of the publication.

The online catalogues for the sources neither gave rights holder information, contact for publication permission, nor indicated the terms and conditions for the use of images.

This example illustrates the point that FAIR data is not necessarily open data, but data with clearly articulated reuse conditions. Notice that the problem here was not openness in the first place but a lack of transparency and proper data management that, in originating from external data providers, is out of the control of the researcher community. If the longevity of cultural heritage data is defined by their presence in scientific, cultural, and social discourses, then once we lose access to their reuse conditions, we lose them entirely.

---

17 *Cendari*, http://www.cendari.eu/
18 'Publicly Available Research Guides', *Cendari*, http://www.cendari.eu/thematic-research-guides/available-research-guides

The Risk of Losing the *Thick Description* upon the Remediation of Cultural Heritage

The advent of digital research infrastructures opened up a radically new frontier for the interactions with cultural heritage of both scholars and the public in an increasingly data-intensive and collaborative research ecosystem. As an active response to the impact of the digital age on scholarly and archival practice, a range of research data aggregation and discovery projects of different scopes and sizes have been created, such as: Europeana Collections,[19] IPERION CH,[20] and CENDARI.[21] They all have the mission to build bridges, interlinks, and networks (e.g., co-referencing systems, conceptual models, ontologies, semantic web frameworks) across different types of resources and institutions in order to enable the browsing of this heterogeneous content within a single search and discovery space. Although many of these infrastructures are facing sustainability challenges, their role in computationally-enhanced scholarly workflows is indispensable. Leveraging the power of big data and linked data approaches enables scholars to gain access to cultural heritage resources across institutional and national boundaries, and to explore new, macro-level perspectives and connections between distant events, communities, or traditions that could not have been made visible via traditional manual methods.

In addition to opening up new paradigms and epistemic models of knowledge creation, such research infrastructure initiatives also should be credited with having played a catalytic role in the development, promotion, and implementation of shared protocols and standards (like the Linked Open Data paradigm in arts and humanities)[22] to guarantee interoperability between heterogeneous data resources. Papers that report on data collection procedures for the research infrastructure projects EHRI (European Holocaust Research Infrastructure)[23] and

---

19 'Europeana Collections', *Europeana Collections*, https://www.europeana.eu/portal/?locale=en
20 'Iperion Homepage', *Iperion CH,* http://www.iperionch.eu/
21 *Cendari,* http://www.cendari.eu/
22 *Linked Data — Connect Distributed Data across the Web,* http://linkeddata.org/
23 Mike Bryant et al., 'The EHRI Project — Virtual Collections Revisited', in *Social Informatics,* ed. by Luca Maria Aiello and Daniel McFarland (Cham, Switzerland: Springer International Publishing, 2015), pp. 294–303, https://doi.org/10.1007/978-3-319-15168-7_37

CENDARI[24] provide an insight into the various challenges the participating projects and institutes had to face, as well as into the, sometimes, herculean efforts they made to put their records onto the world map of computationally remediated digital horizons.

Here, again, the standardisation of shared metadata has brought not only technical and financial challenges, but also epistemological challenges: the new ways in which cultural resources have been made available as a part of global networks affects the systems of discovery and knowledge creation. Following up on, and investigating the changing archival practices of cultural heritage institutions in the age of big data, the aforementioned KPLEX project[25] uncovered many important epistemological implications for the computational turn.

One of these has to do with losing control over the remediated records of archival knowledge and its complexity. In the course of traditional interactions, such as in-person visits or one-on-one consultations, archivists had the possibility of freely guiding the researcher through the collections and transferring all relevant knowledge to the specific research question. Since such mutual exchange-driven means of discovery are not possible in a computationally mediated context, researchers are left alone with the task of interpreting the specific datasets that had been harvested from institutions. Practitioners' concerns about misinterpretations and misuse of the data they had carefully curated were clearly and repeatedly indicated in the interviews.[26]

A speciality of data management in arts and humanities, therefore, is that it is highly dependent on external data providers, that is, the cultural heritage institutions.[27] As was also touched on in the CENDARI case study above, due to this dependence, certain aspects of data management and FAIRification efforts remain out of the control of researchers. In addition, the ways in which cultural heritage materials are made available to them define and, in many cases, impose limitations on the accessibility of complex knowledge structures. As a result of the separation of data from its context of creation (i.e. from the institution,

---

24   Jakub Beneš et al., *The CENDARI White Book of Archives* (2016), http://www.cendari.eu/sites/default/files/WhiteBook-Web.pdf
25   *KPLEX*, www.kplex-project.eu
26   Priddy and Horsley, 'Deliverable D3.1 Report' pp. 52–53, 64–68.
27   However, arts and humanities are not the only disciplines that are dependent on external data providers, see, for example, medical and health care studies.

its curators, and its wider provenance), collection descriptions that are part of the standardised and aggregated metadata remain the only reference points for the long history of records.

Creating descriptions is, therefore, a pivotal process, but also a complex task. Practitioners showed an awareness of how much the preparation of these online representations, and the alignment of the richest possible descriptions with their limited space and capacity, is an interpretative practice. As has also been pointed out by Wendy M. Duff and Verne Harris,[28] personal decisions made in the course of this knowledge transfer are inherently biased and will, therefore, foreground certain pieces of information, while leaving others sunk in analogue practices and tacit knowledge.[29] One thing, however, is clear: the separation of the data from the curators who bear this knowledge, instead providing an impoverished form of online access to such remediated knowledge representations, necessarily leads both to limitations in conveying their complexity and to simulacra that are misleading in their apparent completeness. This is crucial, because the loss of information is the loss of the continuous narratives of the origins and subsequent treatment of a source, which is critical to interpreting how it might be used in relation to other research sources — a central technique by which historical interpretations are corroborated and verified.

Consequently, the loss of this knowledge complexity imparts serious deficits in the reuse and interoperability potential of data made openly available by the hard work of curators, just as it may impoverish researchers' interpretation and understanding of the possible uses of sources. In other words, hiddenness and the loss of the *thick descriptions*[30] of holdings is a part of the process of making

---

28 Wendy M. Duff and Verne Harris, 'Stories and Names: Archival Description as Narrating Records and Constructing Meanings', *Archival Science*, 2.3 (2002), 263–85, https://doi.org/10.1007/BF02435625

29 This typically involves not only dynamics of foregrounding and backgrounding but also changes in scope and detail. 'Changing practice therefore carries risks of skimming over knowledge complexity to produce a simulacrum that represents less of an item's deviation from the collection in which it has been placed. In this way, differences between collections may become exaggerated as practitioners' "closeness" reinforces the unique value and identity of a collection as the smallest unit in their purview, while the complexity that distinguishes the unique value of items may be hidden.' Priddy and Horsley, 'Deliverable D3.1 Report', p. 83.

30 The term *thick description* is borrowed from cultural anthropology, a prominent subfield of the study of cultural heritage. The term was coined by the twentieth-century

historical and cultural records available for digital and computational discovery. Researchers in the arts and humanities always need multiple sources to verify interpretations, but this requires a deep knowledge of source provenance. Therefore, without complexity and context, the FAIR principles of maximum reusability and interoperability cannot be achieved on an epistemic level, even if they can be achieved technically.

As the results of the aforementioned Europeana survey suggest, the *thick description* of holdings is not the only layer of archival knowledge that might remain invisible or lost in a computationally mediated context of discovery. Practitioners' concerns about the non-digitised or offline substructure of an iceberg of knowledge, with the levels invisible below the water being forgotten and 'buried at deeper levels of accessibility during this transitional period' were clearly articulated in the KPLEX interviews.[31] It is a serious threat that a new generation of scholars might lose this awareness of materials and knowledge structures that have submerged beyond the digital horizon, resulting in a situation where one has to know what it is one cannot find. The main danger of this effect is that it may skew research towards what is easily available, easy to find, and, ideally, available freely online. This would generate a further enrichment and even greater visibility of this yet very small fraction of cultural heritage. Such asymmetry and distortion can cause potentially irreparable damage to our understanding of human culture. As Jennifer Edmond points out in her 2015 study, such distortion effects are also arising from the fact that, contrary to the essentially transnational nature of historical research, the digitisation of cultural heritage has largely been funded, and continues to be funded, along national lines, and not every country or institution has access to the same resources.[32] This results in substantial differences in the digital and online footprint

---

philosopher Gilbert Ryle (1900–1976), but it was the anthropologist Clifford Geertz who developed the concept into an ethnomethodological key notion with sufficient explanatory power, in his seminal work *The Interpretation of Cultures* (Clifford Geertz, *The Interpretation Of Cultures*, rev. ed. (New York: Basic Books, 2000), pp. 9–10). Geertz described the practice of *thick description* as a way of providing cultural context and meaning that people place on actions, words, things, etc. *Thick descriptions* provide enough context so that a person outside the culture can make meaning of the behaviour. Since then, the term and the methodology it represents has gained currency in the social sciences and beyond, and so today, *thick description* is used in a variety of fields of cultural study.

31 Priddy and Horsley, 'Deliverable D3.1 Report', p. 79.
32 Edmond, 'Tradition and Innovation', pp. 2–9.

of the various institutional holdings: wealthier institutions might have a stronger representation and, therefore, impact on historical research than those who have limited access to funding. This, in turn, 'risks creating perverse incentives for historians that bring to mind the tale of the drunk looking for his lost keys under the lamppost — not because that is where they were lost, but because that is where the light is'.[33]

Amid FAIRification efforts, as we develop our knowledge creation ecosystem to the next level — from a human-scaled to a machine-actionable one — the lessons that can be learned from these insights are crucial, and not only for researchers in the arts and humanities. Being attentive, along with maintaining an attitude of critical reflection regarding overall progress and limited or immature cases of openness, may help identify phenomena and situations where the principles enshrined in the first two letters of FAIR, 'findability' and 'accessibility', come into conflict with the last letter, 'reusability'. If we want to play it right in the computational research ecosystem, the ability to recognise and amend such contradictions is an essential skill for all researchers and in all research practices. Allowing knowledge icebergs and *thick descriptions* to remain invisible beyond the digital horizon would be an unreasonable price to pay for the sake of a paradigm shift. Being aware of them is a guarantee that we will not have to pay this price and can realise the promises of the innovative revolution to the full, thus enabling new forms of scholarly insight and communication.

The Scholarly Data Continuum

The previous sections highlighted that, in contrast to the hard sciences, the initial data in the arts and humanities is *collected*[34] rather than *generated*,[35]

---

33 Ibid., p. 4.
34 This distinction and its epistemological consequences are also articulated in Johanna Drucker's study on *capta* versus *data* where capta is 'taken' (the term *capta* stems from the Latin word for 'to take'), constructed, and is rooted in the co-dependent relation between the observer and the experience, while data represents observer-independent models of knowledge given as a natural representation of pre-existing fact. See Johanna Drucker, 'Humanities Approaches to Graphical Display', *Digital Humanities Quarterly*, 5.1 (2011), http://digitalhumanities.org/dhq/vol/5/1/000091/000091.html
35 Claudine Moulin et al., *Research Infrastructures in the Digital Humanities* (Strasbourg: European Science Foundation, 2011), p. 5, http://www.esf.org/fileadmin/user_upload/esf/RI_DigitalHumanities_B42_2011.pdf

and thus the digitisation of cultural heritage is an indispensable base for research in these disciplines. However, considering the highly intertwined systems of knowledge representation and knowledge creation[36] — a phenomenon that is commonly referred to in arts and humanities discourse as the illusion or oxymoron of raw data[37] — it is rather difficult to decouple this base of cultural data from the layers of analysis built upon them.

Embedded within the practices of making cultural heritage material digitally available, there is a series of decisions cultural heritage curators have to make: they range from decisions on what and what not to preserve, choosing classification systems and metadata schemas, determining the way in which texts and artefacts are photographed; to the ways in which text corpora are transcribed, encoded, or the OCR (optical character recognition) is corrected. All of these decisions impose a perspective, and thus an influence, on our perceptions of, and access to, data within a research environment. The creation of digital objects for arts and humanities research purposes is, therefore, not an innocent practice: it is not merely a prerequisite for digitally-enabled research, but is an important scholarly activity in itself. The initial layer of interpreting, preparing, and pre-processing cultural heritage data is, therefore, provided by the heritage institutions, a process that enables and gives access to other layers of analysis and knowledge creation resulting from scholarly activities.

Within the current practice, these different layers of analysis are separated by institutional silos and only in the rarest cases can they

---

36 See discussion on the 'fuzzy, implicitly highly networked data' in the humanities that questions the separability of the data areas of primary- and intermediate-data-results in Patrick Sahle and Simone Kronenwett, 'Jenseits der Daten: Überlegungen zu Datenzentren für die Geisteswissenschaften am Beispiel des Kölner "Data Center for the Humanities"', *LIBREAS. Library Ideas*, 23 (2013), https://libreas.eu/ausgabe23/09sahle/. Sahle and Kronenwett argue that by digitising the research process, the various types of research data merge into a continuum where narratives and knowledge creation practices are present from the initial data to the research output publications and keeping this continuum together poses special challenges in data management and hosting infrastructure. The challenges in keeping together different mediums of knowledge creation, data and software in the first place is a general and major challenge in sustainable in reproducible data management and is a topic that deserves more detailed discussion than it can receive within the framework of the present paper.

37 Virginia Jackson and Lisa Gitelman, 'Introduction', 'in *Raw Data' Is an Oxymoron*, ed. by Lisa Gitelman, Geoffrey C. Bowker, and Paul N. Edwards (Cambridge, MA: MIT Press, 2013), pp. 1–14, https://doi.org/10.7551/mitpress/9302.003.0002

stay connected with each other. As a result, the actual continuum of the knowledge creation procedures of the cultural heritage domain is barely reflected in its infrastructure and data management practices.

A key recommendation in the FAIR principles, which aims to facilitate access to research data, is that data should be stored in trusted and sustainable digital repositories.[38] Taking the view from the researchers' side of cultural heritage knowledge creation, the landscape of outputs and throughputs show a rather fragmented picture. At the time of writing, the reference repository catalogue re3data.org lists 206 data repositories under the subject label 'humanities'; a relatively small number, not only in comparison with umbrella disciplines with more robust traditions of 'data-drivenness' such as life sciences (1,132 results), but also compared to the sibling disciplinary group, social and behavioural sciences (331 results).[39] The low number of repositories suggests lower demand for data sharing services, or, at least, a less established data sharing culture in the arts and humanities than in other fields of study.[40] On the other hand, however, several recent studies herald an increasing interest in data sharing in the arts and humanities at a global disciplinary scale.[41] For instance, in Ruth Mostern and Marieka Arksey's 2016 study,[42] which surveyed the target users of the Collaborative for Historical Information and Analysis

---

38 Hodson et al., 'Turning Fair DATA into Reality', p. 18.
39 *Re3data Registry of Research Data Repositories*, www.re3data.org
40 In their 2013 study investigating disciplinary differences in data management practices, Katherine G. Akers and Jennifer Doty arrive at similar conclusion. They found that in their university (Emory University) arts and humanities researchers tend not to store their data using university-based servers but instead rely heavily on computer/external hard drives and internet-based storage. Katherine G. Akers and Jennifer Doty, 'Disciplinary Differences in Faculty Research Data Management Practices and Perspectives', *International Journal of Digital Curation*, 8.2 (2013), 5–26 (p. 9), https://doi.org/10.2218/ijdc.v8i2.263
41 Rinke Hoekstra, Paul Groth, and Marat Charlaganov, 'Linkitup: Semantic Publishing of Research Data', in *Semantic Web Evaluation Challenge*, ed. by Valentina Presutti et al., Communications in Computer and Information Science (Cham, Switzerland: Springer International Publishing, 2014), pp. 95–100, https://doi.org/10.1007/978-3-319-12024-9_12; Sandra Collins et al., *Going Digital: Creating Change in the Humanities* (Berlin: ALLEA E-Humanities Working Group Report, 2015), p. 6, https://hal.inria.fr/hal-01154796
42 Ruth Mostern and Marieka Arksey, 'Don't Just Build It, They Probably Won't Come: Data Sharing and the Social Life of Data in the Historical Quantitative Social Sciences', *International Journal of Humanities and Arts Computing*, 10.2 (2016), 205–24, https://doi.org/10.3366/ijhac.2016.0170

(CHIA) database, ninety-four percent of the respondents indicated that they would consider putting their data in a repository.[43]

Understanding this large gap between intentions, real willingness, and practice is a key step towards the development of research data management services and recommendations that match humanities researchers' needs.

Data in Arts and Humanities — Still a Dirty Word?

Sharing data necessarily implies having or owning data. In addition to the aforementioned complexities in the shared ownership of primary sources, which forms a major hindrance to data sharing, having data or working with data is not always a straightforward process, especially in the traditional fields of arts and humanities. Iterated and large-scale surveys would be beneficial for assessing whether, and to what extent, the term 'data' is still a dirty word in the increasingly digital humanities disciplines and how the evolving landscape of open data and FAIR data policies impact and transform such conceptions of data.[44]

Surveys from the past five years[45] reveal a great deal of uncertainty in the arts and humanities researchers' conception of data and its

---

43   This seems significant progress over, for example, Diane Harley et al., *Assessing the Future Landscape of Scholarly Communication: An Exploration of Faculty Values and Needs in Seven Disciplines* (Berkeley, CA: Center for Studies in Higher Education, 2010), https://escholarship.org/uc/cshe_fsc. In this study, evidence is shown that historians are cautious about sharing work publicly until it is well-polished. Similar to many other fields in the arts and humanities, drafts are generally circulated by email among a small network of trusted colleagues for comment, feedback, and improvement. The study also points out how sharing habits are dependent on career stages; while graduate students and pre-tenure scholars may harbour fears that openly shared, in-progress work could be heavily criticised or poached, tenured scholars tend to be more comfortable with sharing early research ideas and other in-progress work. As concerns data sharing, the study argues that 'While scholars have varied opinions regarding the sharing of primary archival data, few scholars share their research notes, databases, or other intermediary interpretations of archival material; those who do usually wait until they have formally published their research' (p. 451).

44   Alicia Hofelich Mohr et al., 'When Data is a Dirty Word: A Survey to Understand Data Management Needs Across Diverse Research Disciplines', *Bulletin of the Association for Information Science and Technology*, 42.1 (2015), 51–53, https://onlinelibrary.wiley.com/doi/full/10.1002/bul2.2015.1720420114

45   Akers and Doty, 'Disciplinary Differences'; Mohr et al., 'When Data is a Dirty Word'; Hélène Prost, Cécile Malleret, and Joachim Schöpfel, 'Hidden Treasures: Opening Data in PhD Dissertations in Social Sciences and Humanities', *Journal of*

applicability to their own work.⁴⁶ Concerns and difficulties around the concept of data were clearly reflected in responses to the survey conducted by Jennifer L. Thoegersen in 2018 and published under the title '"Yeah, I Guess that's Data": Data Practices and Conceptions among Humanities Faculty'.⁴⁷ Here, humanities faculty members from the University of Nebraska-Lincoln were interviewed about their data management practices; all the participants expressed some level of uncertainty while talking about their own data management practices. For example, someone asked, 'Does that sound right?',⁴⁸ after providing a definition of data.

The study does not specify any information about the research practices of the faculty members, so the intriguing question is left open as to whether there is any correlation between data awareness and the level of integration of computational methods into the respective research workflows. Another relevant feature of arts and humanities research that may explain confusion around the notion of data is the great variety in the types of sources and information throughputs and outputs (laser scanner data, musical notations, voice recordings, annotations, critical editions etc.) produced by the wide ranging disciplines that come under the umbrella term of arts and humanities, as well as under the umbrella term data in computational research contexts.

The Critical Mass Challenge and the Social Life of Data

The intensifying discourse around data conceptions and data characteristics clearly indicates a shift in the paradigm towards data-driven and computational methods across the whole disciplinary range of the arts and humanities. Yet, there are still plenty of interrelated

---

Librarianship and Scholarly Communication, 3.2 (2015), http://doi.org/10.7710/2162-3309.1230; Jennifer L. Thoegersen, '"Yeah, I Guess that's Data": Data Practices and Conceptions among Humanities Faculty', *Libraries and the Academy*, 18.3 (2018), 491–504.

46   As Jennifer L. Thoegersen remarks, researchers in arts and humanities may not be comfortable describing their scholarly and academic work as data. A potential reason behind this is that in their data conceptions are tied to the prototypical data representations such as numerical or quantitative description of data. Thoegersen, '"Yeah, I Guess that's Data"', 492.
47   Thoegersen, '"Yeah, I Guess that's Data"'.
48   Ibid., p. 501.

issues that prevent data sharing in subject repositories (which are, as we have seen, central data services in the implementation of the FAIR principles) and hamper reuse in becoming an entrenched and integral part of scholarly practices. In their 2016 paper 'Don't Just Build It, They Probably Won't Come: Data Sharing and the Social Life of Data in the Historical Quantitative Social Sciences', Mostern and Arksey capture many such interrelated problems that define the current repository landscape in the arts and humanities,[49] which lingers in a vicious cycle of data repository failure. They make these observations in the context of quantitative historical research, but it is not a stretch to extend these insights to the multitude of scholarly communities in the arts and humanities, keeping in mind that they are not equally plagued with the problems described.

As has been pointed out in several other discipline-specific data management studies, there is a lack of incentives and rewards to dedicate to the considerable amount of time, effort, and expertise needed to prepare data for computational analysis and make it compliant with the standards and data models of the repositories.[50] Consequently, only a small user community is open to taking steps in sharing data and thus contributing to the development of repositories. As a result, the limited number of contributions coming from this small user base will not attract further communities to visit or contribute to them.[51] In addition,

---

49 Mostern and Arksey, 'Don't Just Build It'.
50 Robin Rice and Jeff Haywood, 'Research Data Management Initiatives at University of Edinburgh', *International Journal of Digital Curation*, 6.2 (2011), 232–44, https://doi.org/10.2218/ijdc.v6i2.199; Alex H. Poole, 'Now is the Future Now? The Urgency of Digital Curation in the Digital Humanities', *Digital Humanities Quarterly*, 7.2 (2013), http://www.digitalhumanities.org/dhq/vol/7/2/000163/000163.html; Catherine Anne Woeber, 'Towards Best Practice in Research Data Management in the Humanities' (unpublished master's dissertation, School of Information Management, Victoria University of Wellington, 2017), http://researcharchive.vuw.ac.nz/handle/10063/6620
51 Note that guaranteeing the presence of a target audience by reaching a critical mass of content was the recipe for success of the two academic sharing and networking platforms ResearchGate and Academia.edu, which even today are commonly used. We can learn a lot from the failures that underlie their conceptual design and what became visible only after they reached a critical level of user engagement. Although the original aim of both platforms was to help researchers go beyond paywalls and increase the availability of their research, the low entry thresholds (direct upload of PDFs, no custom metadata, no licensing options) conserved bad sharing behaviours (low awareness of copyright, which article versions are allowed to be legally shared, low awareness of the importance of licensing issues, support for freemium business models based on selling data on user behaviours) on such

repository developers and standardisation bodies then do not receive a significant enough input foundation from diverse sources that could serve as a sufficient and informative basis for developing infrastructural components (widely accepted metadata standards tailored to specific data types, for example, or analytical tools for opening up the boxes of deposited datasets etc.) such as could truly increase the visibility and discoverability of deposited data, and that could also connect them with other databases or datasets. This lack of momentum preserves the scattered landscape of subject repositories, and also maintains the status of repository users as an invisible or only slightly visible part of the wider disciplinary communities. This prevents their work and approaches from being both accessible and strongly represented to students and peers. In turn, it does not encourage them to share their data; thus, ultimately, the strongest appeal for the use of repositories is not able to work its charm.

Having been inspired by the 2003 study by Jeremy P. Birnholtz and Matthew J. Bietz,[52] Mostern and Arksey describe this complex phenomenon as the lack of the social life of data. Recognising the importance of having a community aspect around robust data sharing culture (wherein documents and deposited datasets are not only a means for delivering information, but are also meant for maintaining social groups and the professional exchange around them), they came to the important conclusion that repositories can only succeed as long as scholarly communities create social communities around them.[53]

---

a massive scale that it seriously slowed down the development and large-scale uptake of more sustainable, transparent, and legal ways of self-archiving (such as the use of preprint servers). For more discussion on such controversies see: Jonathan P. Tennant, 'ResearchGate, Academia.Edu, and Bigger Problems with Scholarly Publishing', *Green Tea and Velociraptors* (2 February 2017), http://fossilsandshit.com/researchgate-academia-edu-and-bigger-problems-with-scholarly-publishing/

52   Jeremy P. Birnholtz and Matthew J. Bietz, 'Data at Work: Supporting Sharing in Science and Engineering', in *Proceedings of the 2003 International ACM SIGGROUP Conference on Supporting Group Work*, GROUP '03, ed. Kjeld Schmidt, Mark Pendergast, Marilyn Tremaine and Carla Simone (New York: ACM, 2003), pp. 339–348, https://doi.org/10.1145/958160.958215

53   These observations show congruency with the main findings of a much earlier study on the uptake and use of digital resources in the arts and humanities, namely the LAIRAH project (Log analysis of Internet Resources in the Arts and Humanities; see a project description in C. Warwick et al., 'Evaluating Digital Humanities Resources: The LAIRAH Project Checklist and the Internet Shakespeare Editions Project', in *Openness in Digital Publishing: Awareness, Discovery, and Access. Proceedings of the 11th International Conference on Electronic Publishing*, Vienna, 13–15

This primarily includes peer evaluation of the deposited datasets. Data peer review is not only a vital step towards the acknowledgement and recognition of research data sharing, but, as their survey shows, it is also important in building user confidence, as seventy percent of historians responding to their survey indicated that a peer review process or citation option as part of the data submission process would increase their incentive to do so.

The idea of providing infrastructural support to bring the scholarly practices of data depositing and data peer review into closer proximity is also expressed in a checklist of recommendations in the Log Analysis of Internet Resources in the Arts and Humanities (LAIRAH) project. According to these recommendations, the ideal digital resource should be as follows:

1. it should have access to good technical support, ideally from a centre of excellence in digital humanities;

2. it should recruit staff who have both subject expertise and knowledge of digital humanities techniques; and

3. it should also retain this expert staff by having constant access to funds.[54]

Data peer review along these lines — that is, focusing on the support and joint development of transparent and good quality data creation without the power dynamics and the gatekeeping function that are causing serious challenges in the institution of the traditional article and book peer review[55] — could also be interpreted as a significant

---

June 2007, ed. by Leslie Chan and Bob Martens (Vienna, Austria: ELPUB, 2007), pp. 297–306, https://publik.tuwien.ac.at/files/pub-ar_7877.pdf). The project was based at UCL's School of Library Archive and Information Studies and was aimed at identifying the various factors (under the categories of content, user, maintenance and dissemination) that influence the long-term sustainability and use of digital resources in the humanities. Reaching a critical mass and gaining prestige within a university were found to be vital in the sustainability and longevity of digital infrastructures. In addition, the importance of good project staff and the availability of technical support have also been pointed out. As a result of the research, Warwick et al. ('Evaluating Digital Humanities') provided a checklist of recommendations to facilitate both the successful design of digital infrastructures and the recognition and culture around them.

54   Warwick et al., 'Evaluating Digital Humanities', pp. 302–03.
55   See, for example, Jonathan P. Tennant, 'The State of the Art in Peer Review', *FEMS Microbiology Letters*, 365.19 (2018), https://doi.org/10.1093/femsle/fny204

contribution to a more sustainable and more inclusive culture of research evaluation in general. At the same time however, the third LAIRAH recommendation stated above also indicates the serious sustainability challenges for such models in terms of funding. The ability to maintain, in repositories, both a technically and disciplinarily highly skilled expert staff, who have the capacity to provide a thorough evaluation of the massive number of data deposits that can be expected as a result of FAIR policies, does not seem to be a viable option. As a potential alternative, institutional data stewards[56] and data centres like the Leiden University Centre for Digital Humanities (LUCDH)[57] could at least partially fulfil this role.

An additional challenge in facilitating the culture of data evaluation in the arts and humanities, as has been pointed out by others, is that the scholarly practice of data peer review is still substantially lagging behind the traditional paradigm of research article publishing, which serves as academia's highest value currency.[58] Bringing these two forms and practices of scholarly communication, data sharing, and article or book publishing, closer to each other is a key step towards a more open, more connected, more transparent, and more sustainable research data management ecosystem.

The Risk of Losing the *Thick Description* — Again

Relying on domain-relevant community standards is critical to avoid having deposited datasets being buried in isolated 'data tombs', and to

---

56 Rec. 13 of the FAIR Data Action Plan (Hodson et al., 'Turning FAIR Data into Reality', p. 73.) recommends developing two cohorts of professionals to support FAIR data: data scientists embedded in those research projects that need them, and data stewards who will ensure the management and curation of FAIR data.
57 Researchers who need help or have questions regarding the critical use of digital technology and computational approaches in disciplines of the humanities can get support from the Leiden University Centre for Digital Humanities (LUCDH). A case study published in a recent collection of FAIR data advanced use cases from the Netherlands gives an insight into how this type of institutional support might work in an arts and humanities context. Melanie Imming, 'FAIR Data Advanced Use Cases: From Principles to Practice in the Netherlands', *Zenodo* (2018), 33–35, https://doi.org/10.5281/zenodo.1246815
58 E.g., Anne Baillot, 'A Certification Model for Digital Scholarly Editions: Towards Peer Review-Based Data Journals in the Humanities', *HAL* (2016), halshs-01392880, https://halshs.archives-ouvertes.fr/halshs-01392880/document

increase the social life of data by making it interoperable and connectible with other data sources. Achieving compliance with metadata standards is a prerequisite for improving the visibility, accessibility, interoperability, and linking of digital resources. Shared standards open up datasets for integration with research across different sectors, provide additional layers of context, and enable research methods that have not been previously available to the humanities.

However, aligning the application and use of repository standards with the long history of data curation cannot always be achieved without making compromises. In some cases, enforcing a commitment to shared standards can lead to a similar loss of detail and information, as was seen in the context of the aggregation of standardised and machine-interoperable metadata from cultural heritage institutions. In their 2014 and 2016 studies, Rinke Hoekstra and his co-authors investigated data sharing practices in the humanities and their compliance with linked discovery context.[59] They identify two cases in which the risk of losing provenance information is especially high.

First, when data is deposited in bigger, discipline-specific data curation projects with top-down standards (such as the North-Atlantic Population Project (NAPP), the Clio Infra repository, or the Mosaic project), Hoekstra et al. point out that the sheer scale of such databases and the top-down fashion of their data curation standards are not always suitable for smaller datasets created by individual researchers. This makes it difficult for them to share their research in a sustainable way.[60]

Second, not every researcher has equal access to the computational resources, expertise, and skills necessary to create and operate a digital data collection. To address this problem a number of low-barrier-to-entry repository data services have been created (e.g., EASY, Dryad, Dataverse, and Figshare). These services are important pillars of scientific data sharing infrastructure as they help to satisfy the growing demand for sustainable data sharing and archiving services. They enable easy data upload in most formats; ensure data is citable via

---

59 Hoekstra, Groth, and Charlaganov, 'Linkitup'; Rinke Hoekstra et al., 'An Ecosystem for Linked Humanities Data', in *The Semantic Web*, ed. by Harald Sack et al., Lecture Notes in Computer Science (Cham, Switzerland: Springer International Publishing, 2016), pp. 425–40.

60 Hoekstra et al., 'An Ecosystem', p. 426.

persistent identifiers, and also guarantee long-term archival storage. On the other hand, as argued in the earlier study, these generic-scope data sharing platforms bear hidden limitations on discoverability and productive reuse.[61] The first limitation is the result of the rather isolated presentation of the data: a landing page is provided for each deposited item, but the items are not embedded into a related network of relevant datasets. This might stem from these services' primary focus on long-term preservation. More importantly, in such low-barrier-to-entry data services, metadata schemas associated with data publications are usually limited to a minimum set of information (authors, title, publication date, free text tags, and categories) and inflexible licensing options that can neither fully cover the complex ownership relations in cultural heritage data, nor are sufficient for providing detailed provenance information.

In both cases we face the minimal common denominator problem: minimally flexible and minimally specified metadata schemas serving as a common base for the accommodation of large amount of heterogeneous data will necessarily bring about at least some loss of information that would otherwise enable productive reuse of the dataset. Such limited possibilities for contextualising and documenting data may keep important assumptions, procedures, processes, and decisions that were made at the different stages of data collection and curation hidden from potential re-users of the deposited dataset. As Carlson and Anderson remind us, data are always cooked in specialised ways within each and every research project.[62] Making the steps of this cookery process explicit is especially important when data designed to answer specific research questions are derived from cultural artefacts carrying their own long life-stories and *thick descriptions*.

Recognising these limitations, which are imposed by insufficient metadata and deficient documentation on reuse, highlights an important aspect of successful data management. That is, to make datasets truly reusable, data should achieve autonomy from their curator. In Carlson and Anderson's words: 'Data re-use not only involves the disconnection of data from the people they represent but also from the researchers

---

61   Hoekstra, Groth, and Charlaganov, 'Linkitup', p. 96.
62   Carlson and Anderson, 'What *Are* Data?', 144; also cited by Poole, 'Now is the Future Now?', para. 20.

who collected them. This opens up the central question as to how data collected or constructed by one researcher can be trusted or even understood by another.'[63]

In the arts and humanities this act of disconnection is a recurring pattern. Artefacts first become separated from their producers (e.g. from the photographer or writer) when they are brought into cultural heritage institutions. The second separation occurs when digital surrogates, descriptions, and other additions to the history, discoverability, and *thick description* of artefacts — in optimal cases at least — step outside the bounds of the cultural heritage institutions responsible for their preservation and digital curation. The third separation occurs when research data derived from these digitally available cultural data is shared and reused, making it available for continuous enrichment and analysis in multiple research contexts. This third separation is a slowly emerging scholarly practice that is facing many economic, technical, institutional, infrastructural, but primarily, and most importantly, cultural barriers. The more support data sharing practices receive, the more important the question is of how to keep these multiple contexts of the *thick descriptions* of cultural data available for continuous analysis and enrichment. Enabling FAIR data management to realise its promises in the arts and humanities requires a mutual understanding between the epistemic cultures of the various stakeholders involved in the co-creation of the scholarly data continuum, ranging from primary sources to multiple reuse cases.

Conclusions: On our Way towards a Truly FAIR Ecosystem for the Arts and Humanities

It is now beyond question that opening up access to scholarly knowledge is a key value of twenty-first-century academia. The paradigm shift towards digital and computational research methods brings about more sustainable, more connected, and community-driven models of scholarly production. Global policies like FAIR data management have a vital role in catalysing and streamlining such innovations, and also in transposing and defining the ways in which research is designed,

---

63 Carlson and Anderson, 'What *Are* Data?', 643.

performed and evaluated, and the ways in which knowledge is shared. However, in order to embrace the new potentials of computational innovation and to implement high-level principles in a way that will serve the flourishing of the arts and humanities disciplines, there are concerns we need to systematically address first, using focussed activities both from within arts and humanities research, and at the level of open science policies. These include:

1. Data-drivenness is not yet a mature concept in the arts and humanities. Consequently, there is a need for consolidated interpretative frameworks aimed at helping to reach consensus about what can be considered to be research data[64] in the arts and humanities disciplines, and what is not. Furthermore, enhancing data literacy requires the integration of new skills and new professional roles with the arts and humanities higher education curricula.

   On the one hand, the institutional availability of expert data curator staff (librarians, data scientists, and digital humanities experts) who have both subject expertise,[65] and knowledge of digital humanities and data science techniques, is critical for the support of the vernacularisation of FAIR data management skills. On the other hand, we can expect that arts and humanities research institutions will not have equal access to these support services, or will not be ready for their rapid implementation. Therefore, as a more flexible and more inclusive solution, we recommend European research infrastructures complement the efforts of research institutions with widely accessible data management services (such as repository finders)[66]

---

64 At the same time, we can expect that the en masse application of global FAIR data policies will also have an incremental and large-scale effect on the notion of data in the arts and humanities as researchers will be forced to interpret certain outputs of their research projects as data.

65 Subject expertise and capacity for one-to-one consultancy would be key contributions for aligning disciplinary culture with data management best practices. This could prevent FAIR from being realised merely as a compulsory administrative task of filling in data management templates tailored to the taste of the different funding bodies, or reducing it to a set of technical requirements.

66 The Data Deposit Recommendation Service (DDRS), which has been developed as functional demonstrator within the Humanities at Scale project, an offspring of DARIAH-EU, is a good example of services helping to establish good data

and advocacy activities (webinars, workshops, e-learning materials, collecting, and sharing exemplary case studies). For instance, the translation of science policies (which are often expressed in science-centric language) into widely applicable terms and disciplinary contexts is an important step in preventing humanities researchers from feeling marginalised and disengaged. By uncovering some of the cornerstones for reconciling disciplinary traditions with FAIR data management, this chapter aims to contribute to this translation.

2. In the arts and humanities, data are collected rather than generated. The history of practices determines the way in which cultural resources are made available. Dealing with non-digital heterogeneous materials has many implications for data fluidity and data-reuse.[67] Most importantly, being attentive to knowledge structures submerged beyond the digital horizon is essential, if we are to avoid research being skewed towards easily available, easy to find online resources, generating further enrichment and even greater visibility — but only for this very small fraction of cultural heritage. Such asymmetry and distortion can cause potentially irreparable damage to our understanding of human culture. Building research infrastructures that do not completely isolate data from their source institutions, but rather incorporate traditional archival practices and knowledge, and facilitate mediation and connections between the computational and the analogue epistemic cultures, could help avoid such potential distortions.

3. In the arts and humanities, data show a highly networked but also highly scattered picture. They are networked in the sense that, due to the intertwined systems of knowledge representation and knowledge creation, it is rather difficult to decouple the never-raw source data from the layers of

---

management practices in arts and humanities. *DDRS*, https://ddrs-dev.dariah.eu/ddrs/

67 Anne Baillot, Michael Mertens, and Laurent Romary, 'Data Fluidity in DARIAH — Pushing the Agenda Forward', *BIBLIOTHEK Forschung Und Praxis*, 39.3 (2015), 350–57, https://doi.org/10.1515/bfp-2016-0039

analysis that have been built upon them. As a result, scholarly data forms a continuum with not always clearly delineable primary-, intermediate-, and result-data components. In current practice, these different layers of analysis are separated by institutional silos, and only in the rarest of cases can they stay connected to each other. Ensuring that this long continuum is kept together from either end poses special challenges in a data management and hosting infrastructure. Establishing a framework that could serve as a general baseline for interactions between scholars, data centres, and heritage institutions will be an essential component of the FAIR data ecosystem in the arts and humanities domain. Such a trusted network of stakeholders could enable all the relevant actors to connect and together improve access to cultural heritage data, making transactions related to the scholarly use of cultural heritage data more visible and transparent.

4. An important feature of computationally mediated research ecosystems is the autonomy of datasets: as shared assets on a technical level, datasets become disconnected from their creators and contexts of creation, yet epistemologically they still remain, to a certain extent, dependent on these creators and contexts of creation. In the arts and humanities, this act of disconnection is a recurring pattern, and ranges from artefacts first becoming separated from their producers through the opening up of cultural heritage (source) data curated by cultural heritage institutions, to sharing research data and making it available for reuse and reanalysis in multiple research contexts. Such multiple separation events have implications not only in terms of the shared ownership of data, but also in terms of knowledge transfer between these different stakeholder groups. As can be seen, there is a critically high risk of losing contextual information around research sources, which is essential for their productive reuse in the course of remediation of scholarly data. The more support data sharing practices receive, the more important the question: how to prevent this loss and how to keep these multiple contexts of the *thick descriptions* of cultural data available for continuous analysis and enrichment?

Enabling FAIR data management to realise its promise in the arts and humanities requires mutual understanding between the epistemic cultures involved in the co-creation of the scholarly data continuum, ranging from the primary sources to multiple reuse cases. Creating a common online environment to support smooth, end-to-end communication between key actors involved in cultural heritage knowledge creation (cultural heritage institutions, data centres, research institutions, individual researchers) where information on the datasets could be published both manually and automatically (e.g., licensing, citation, reuse, enrichments, and contact information for the persons responsible for curation) would be a key step in keeping together the different layers of analysis, and achieving a better alignment of data creation and curation with downstream reuse.

5. Finally, it is rather difficult to have a fair view of findable, accessible, interoperable, and reusable data management in the humanities without considering the actual situation in the domain of publications. Aligning the slowly emerging scholarly practice of data sharing with the inadequately ageing institutions of book and article publishing is a key step towards a more open, more connected, more transparent, and more sustainable research ecosystem.

Such considerations may pave the way to a better understanding of the discipline-specific challenges in data production and may, therefore, help to realise the promises of the FAIR guidelines in an arts and humanities context. Building a domain-specific data sharing ecosystem will require continuous checks on where the gaps are between the different epistemic cultures, what is hidden, and what remains unknown. Only this can guarantee a truly functioning and sustainable FAIRness, where neither the sunken substructure of the knowledge iceberg, nor the *thick descriptions*, will be lost for good.

Bibliography

Akers, Katherine G., and Jennifer Doty, 'Disciplinary Differences in Faculty Research Data Management Practices and Perspectives', *International Journal of Digital Curation*, 8 (2013), 5–26, https://doi.org/10.2218/ijdc.v8i2.263

Australian FAIR Access Working Group, *Policy Statement on FAIR Access to Australia's Research Outputs*, https://www.fair-access.net.au/fair-statement

Baillot, Anne, 'A Certification Model for Digital Scholarly Editions: Towards Peer Review-Based Data Journals in the Humanities', *HAL* (2016), halshs-01392880, https://halshs.archives-ouvertes.fr/halshs-01392880/document

Baillot, Anne, Laurent Romary, and Michael Mertens, 'Data Fluidity in DARIAH — Pushing the Agenda Forward', *BIBLIOTHEK Forschung Und Praxis*, 39 (2015), 350–57, https://doi.org/10.1515/bfp-2016-0039

Beneš, Jakub, et al., *The CENDARI White Book of Archives* (2016), http://www.cendari.eu/sites/default/files/WhiteBook-Web.pdf

Birnholtz, Jeremy P., and Matthew J. Bietz, 'Data at Work: Supporting Sharing in Science and Engineering', in *Proceedings of the 2003 International ACM SIGGROUP Conference on Supporting Group Work*, GROUP '03 ed. Kjeld Schmidt, Mark Pendergast, Marilyn Tremaine and Carla Simone (New York: ACM, 2003), pp. 339–48, https://doi.org/10.1145/958160.958215

Bryant, Mike, et al., 'The EHRI Project — Virtual Collections Revisited', in *Social Informatics*, ed. by Luca Maria Aiello and Daniel McFarland (Cham, Switzerland: Springer International Publishing, 2015), pp. 294–303, https://doi.org/10.1007/978-3-319-15168-7_37

Carlson, Samuelle, and Ben Anderson, 'What *Are* Data? The Many Kinds of Data and their Implications for Data Re-Use', *Journal of Computer-Mediated Communication*, 12 (2007), 635–51, https://doi.org/10.1111/j.1083-6101.2007.00342.x

Cendari, http://www.cendari.eu/

'Publicly Available Research Guides', *Cendari*, http://www.cendari.eu/thematic-research-guides/available-research-guides

Collins, Sandra, et al., *Going Digital: Creating Change in the Humanities* (Berlin: ALLEA E-Humanities Working Group Report, 2015).

DDRS, https://ddrs-dev.dariah.eu/ddrs/

Drucker, Johanna, 'Humanities Approaches to Graphical Display', *Digital Humanities Quarterly*, 5.1 (2011), http://www.digitalhumanities.org/dhq/vol/5/1/000091/000091.html

Duff, Wendy M., and Verne Harris, 'Stories and Names: Archival Description as Narrating Records and Constructing Meanings', *Archival Science*, 2 (2002), 263–85, https://doi.org/10.1007/BF02435625

Edmond, Jennifer, 'Tradition and Innovation in the Cendari Research Infrastructure', *Review of the National Center for Digitization 25*, ed. by Zoran Ognjanović (Belgrade: Faculty of Mathematics, University of Belgrade, 2015), pp. 2–9.

'Europeana Collections', *Europeana Collections*, https://www.europeana.eu/portal/?locale=en

European Commission, *Digitisation, Online Accessibility and Digital Preservation. Report on the Implementation of Commission Recommendation 2011/711/EU* (2013–2015), http://ec.europa.eu/information_society/newsroom/image/document/2016-43/2013-2015_progress_report_18528.pdf

European Commission, Directorate-General for Research & Innovation, *H2020 Programme Guidelines on FAIR Data Management in Horizon 2020* (26 July 2016), http://ec.europa.eu/research/participants/data/ref/h2020/grants_manual/hi/oa_pilot/h2020-hi-oa-data-mgt_en.pdf

'Guiding Principles for Findable, Accessible, Interoperable and Re-Usable Data Publishing Version B1.0', *FORCE11* (2014), https://www.force11.org/fairprinciples

Geertz, Clifford, *The Interpretation of Cultures*, rev. ed. (New York: Basic Books, 2000).

Harley, Diane, et al., *Assessing the Future Landscape of Scholarly Communication: An Exploration of Faculty Values and Needs in Seven Disciplines* (Berkeley, CA: Center for Studies in Higher Education, 2010), https://escholarship.org/uc/item/15x7385g

Henderson, Margaret E., *Data Management: A Practical Guide for Librarians* (Lanham, MD: Rowman & Littlefield, 2016).

Hodson, Simon, et al., 'Turning FAIR Data into Reality: Interim Report from the European Commission Expert Group on FAIR Data', *Zenodo* (2018), https://doi.org/10.5281/zenodo.1285272

Hoekstra, Rinke, et al., 'An Ecosystem for Linked Humanities Data', in *The Semantic Web*, ed. by Harald Sack, Giuseppe Rizzo, Nadine Steinmetz, Dunja Mladenić, Sören Auer, and Christoph Lange, Lecture Notes in Computer Science (Cham, Switzerland: Springer International Publishing, 2016), pp. 425–40.

Daley, Beth, ed., *Transforming the World with Culture: Next Steps on Increasing the Use of Digital Cultural Heritage in Research, Education, Tourism and the Creative Industries* (The Hague: Europeana Foundation, 2015), https://pro.europeana.eu/files/Europeana_Professional/Publications/Europeana%20Presidencies%20White%20Paper.pdf

Hoekstra, Rinke, Paul Groth, and Marat Charlaganov, 'Linkitup: Semantic Publishing of Research Data', in *Semantic Web Evaluation Challenge*, ed. by Valentina Presutti, Milan Stankovic, Erik Cambria, Iván Cantador, Angelo Di Iorio, Tommaso Di Noia, et al., Communications in Computer and Information Science (Cham Switzerland: Springer International Publishing, 2014), pp. 95–100, https://doi.org/10.1007/978-3-319-12024-9_12

Imming, Melanie, 'FAIR Data Advanced Use Cases: From Principles to Practice in the Netherlands', *Zenodo* (2018), https://doi.org/10.5281/zenodo.1246815

'Iperion Homepage', *Iperion CH*, http://www.iperionch.eu/

Jackson, Virginia, and Lisa Gitelman, 'Introduction', in *'Raw Data' Is an Oxymoron*, ed. by Lisa Gitelman, Geoffrey C. Bowker, and Paul N. Edwards (Cambridge, MA: MIT Press, 2013), pp. 1–14, https://doi.org/10.7551/mitpress/9302.003.0002

Jointly Designing a Data FAIRPORT, Workshop at Lorentz Center@Snellius, Leiden, 13–16 January 2014, https://www.lorentzcenter.nl/lc/web/2014/602/info.php3?wsid=602

Linked Data — Connect Distributed Data across the Web, http://linkeddata.org/

Mohr, Alicia Hofelich, et al., 'When Data is a Dirty Word: A Survey to Understand Data Management Needs Across Diverse Research Disciplines', *Bulletin of the Association for Information Science and Technology*, 42 (2015), 51–53, https://onlinelibrary.wiley.com/doi/full/10.1002/bul2.2015.1720420114

Mostern, Ruth, and Marieka Arksey, 'Don't Just Build It, They Probably Won't Come: Data Sharing and the Social Life of Data in the Historical Quantitative Social Sciences', *International Journal of Humanities and Arts Computing*, 10 (2016), 205–24, https://doi.org/10.3366/ijhac.2016.0170

Moulin, Claudine, et al., *Research Infrastructures in the Digital Humanities* (Strasbourg: European Science Foundation, 2011), http://www.esf.org/fileadmin/user_upload/esf/RI_DigitalHumanities_B42_2011.pdf

Nauta, Gerhard Jan, and Wietske van den Heuvel, *Survey Report on Digitisation in European Cultural Heritage Institutions 2015* (The Hague: DEN Foundation/Europeana/ENUMERATE, 2015), http://enumeratedataplatform.digibis.com/reports/survey-report-on-digitisation-in-european-cultural-heritage-institutions-2015/detail

Poole, Alex H., 'Now is the Future Now? The Urgency of Digital Curation in the Digital Humanities', *Digital Humanities Quarterly*, 7.2 (2013), http://www.digitalhumanities.org/dhq/vol/7/2/000163/000163.html

Priddy, Mike, and Nicola Horsley, 'Deliverable D3.1 Report on Historical Data as Sources', *KPLEX* (2018), https://kplexproject.files.wordpress.com/2018/06/kplex_deliverable-d3-1.pdf

Prost, Hélène, Cécile Malleret, and Joachim Schöpfel, 'Hidden Treasures: Opening Data in PhD Dissertations in Social Sciences and Humanities',

Journal of Librarianship and Scholarly Communication, 3 (2015), https://doi.org/10.7710/2162-3309.1230

Re3data Registry of Research Data Repositories, www.re3data.org

Rice, Robin, and Jeff Haywood, 'Research Data Management Initiatives at University of Edinburgh', *International Journal of Digital Curation*, 6 (2011), 232–44 https://doi.org/10.2218/ijdc.v6i2.199

Rights Statements for in Copyright Objects, http://rightsstatements.org/en/

Sahle, Patrick, and Simone Kronenwett, 'Jenseits der Daten: Überlegungen zu Datenzentren für die Geisteswissenschaften am Beispiel des Kölner "Data Center for the Humanities"', *LIBREAS. Library Ideas*, 23 (2013), https://libreas.eu/ausgabe23/09sahle/

Schöch, Christof, 'Big? Smart? Clean? Messy? Data in the Humanities', *Journal of Digital Humanities*, 2.3 (2013), http://journalofdigitalhumanities.org/2-3/big-smart-clean-messy-data-in-the-humanities/

Tennant, Jonathan P., 'ResearchGate, Academia.Edu, and Bigger Problems with Scholarly Publishing', *Green Tea and Velociraptors* (2 February 2017), http://fossilsandshit.com/researchgate-academia-edu-and-bigger-problems-with-scholarly-publishing/

—— 'The State of the Art in Peer Review', *FEMS Microbiology Letters*, 365.19 (2018), https://doi.org/10.1093/femsle/fny204

Thoegersen, Jennifer L., '"Yeah, I Guess that's Data": Data Practices and Conceptions among Humanities Faculty', *Libraries and the Academy*, 18 (2018), 491–504.

Warwick, C., et al., 'Evaluating Digital Humanities Resources: The LAIRAH Project Checklist and the Internet Shakespeare Editions Project', in *Openness in Digital Publishing: Awareness, Discovery, and Access. Proceedings of the 11th International Conference on Electronic Publishing*, Vienna, 13–15 June 2007, ed. by Leslie Chan and Bob Martens (Vienna, Austria: ELPUB, 2007), pp. 297–306, https://publik.tuwien.ac.at/files/pub-ar_7877.pdf

Wilkinson, Mark D., et al., 'The FAIR Guiding Principles for Scientific Data Management and Stewardship', *Scientific Data*, 3 (2016), https://doi.org/10.1038/sdata.2016.18

Woeber, Catherine Anne, 'Towards Best Practice in Research Data Management in the Humanities' (unpublished master's dissertation, School of Information Management, Victoria University of Wellington, 2017), http://researcharchive.vuw.ac.nz/handle/10063/6620

Index

academia 13, 23, 26, 27, 28, 30, 38, 42, 84, 85, 91, 107, 147, 148, 179, 185, 191, 238, 253, 255, 258. *See also* alternative-academic (alt-ac) career
Academia.edu 32, 36, 186, 202, 252
Academic Book of the Future, The 50
accessibility. *See* FAIR ('findable, accessible, interoperable, and reusable')
Adventure Rock 92
Agile Software Development 141
Alliance of Digital Humanities Organizations (ADHO) 178
alt-ac. *See* alternative-academic (alt-ac) career
alternative-academic (alt-ac) career 72, 88, 147, 227
altmetric. *See* metric: alternative metric (altmetric)
American Council of Learned Societies (ACLS) 165, 216
   ACLS Commission on Cyberinfrastructure in the Humanities and Social Sciences 165
   Our Cultural Commonwealth 216
American Historical Association (AHA) 59, 60, 63, 114
   'Guidelines for the Professional Evaluation of Digital Scholarship by Historians' 114
Ancient Lives 94
Anderson, Ben 238, 257

Anderson, Sheila 229
Andrews, Tara L. 132
Ansax-l (mailing list) 192
Ansaxnet (concept of mailing list for Anglo-Saxonists) 192, 195, 196, 199
AnthroProject 238
Antonijević, Smiljana 5, 132
   Amongst Digital Humanists 5
archiving 12, 26, 27, 37, 40, 41, 42, 66, 71, 74, 92, 240, 243, 244, 246, 250, 256, 257, 260
   self-archiving 12, 253
Aristotle 154
Arksey, Marieka 249, 252, 253
Arts and Humanities Data Service (AHDS) 217, 218
Arts and Humanities Research Council (AHRC) 84, 98
audience 3, 16, 26, 30, 33, 34, 43, 59, 60, 62, 63, 65, 74, 83, 139, 149, 165, 211, 252
authority 4, 6, 9, 11, 22, 39, 41, 42, 51, 64, 66, 68, 73, 105, 111, 115, 116, 117, 145
authorship 13, 27, 35, 36, 110, 116, 137. *See also* registration
   redefinition of, 35
autoethnography. *See* ethnography: autoethnography

Bakhshi, Hasan 84
Bell, Joshua 105, 107
Berners-Lee, Tim 183

Berry, David 123, 128
Bietz, Matthew J. 253
Birnholtz, Jeremy P. 253
BITNET 170, 187
'black box' 15, 40, 123, 124, 127, 128, 131, 143, 150, 153, 215
Blackwell's 178, 184
Blake Archive 87
blog, blogging 2, 4, 9, 10, 15, 31, 32, 34, 59, 61, 65, 187, 198
Bobley, Brett 106
Böll, Heinrich 92
  Cologne Edition 92
Borgman, Christine 73
Brennan, Sheila 118
Brill 74
British Broadcasting Company (BBC) 92, 93
  BBC 2 93
  BBC 4 93
  BBC Radio 4 93
Brunel University 92
Bryn Mawr Classical Review 190
Burgess, Helen 148
Burns, Robert 93
  Auld Lang Syne 93
Burrows, John 168

Cambridge University Press 184
Campanario, Miguel 107, 108
Carlson, Samuelle 238, 257
CARMEN Virtual Laboratory (VL) 69
CASA festival 93
Cavanagh, Sheila 115
Ceci, Stephen J. 108
CENDARI. *See* Collaborative European Digital Archive Infrastructure (CENDARI)
Centre for Robert Burns Studies 92
Centre Virtuel de la Connaissance sur l'Europe (CVCE) 72
Cerquiglini, Bernard 126
certification 26, 27, 38, 39, 40, 42, 58, 70, 71, 173, 199. *See also* metric; *See also* peer review

Church of England 96
CIBER 56, 72
Cicero 133
  De Inventione 133
citation. *See* metric: citation
citizen science 4, 227
Clergy of the Church of England database (CCEd) 96
Clio Infra 256
codework 19, 124, 125, 126, 130, 131, 132, 133, 135, 137, 138, 140, 143, 146, 147, 148, 149, 150, 151, 152, 153, 154, 155, 156. *See also* 'black box'
  as epistemic practice 125, 126, 130, 132, 133, 150, 153, 154, 156
  as memory system 143, 150
  poetics of, 142, 152
coding 8, 14, 15, 124, 125, 132, 133, 134, 135, 136, 137, 138, 139, 141, 142, 143, 145, 148, 149, 150, 151, 154, 155, 156. *See also* 'black box'
collaboration 12, 13, 14, 24, 69, 72, 92, 95, 99, 151, 156, 167, 168, 195
  co-authorship 13
  co-creation 13
  impact of, 13
Collaborative European Digital Archive Infrastructure (CENDARI) 72, 242, 243, 244
Collaborative for Historical Information and Analysis (CHIA) 249
collective intelligence 213, 222
CommentPress 68, 71
Common Languages Resources and Technology Infrastructure (CLARIN) 218
Companion to Digital Humanities, A 178
Computers and the Humanities 166, 169, 174
Conner, Patrick W. 192, 193, 195, 196
copyright 17, 18, 240, 241, 252. *See also* legal liability
Coyne, Richard 128

Craig, Hugh 168
Crossick, Geoffrey 53, 54
    Monographs and Open Access 53
cultural heritage 74, 81, 82, 83, 96, 98, 99, 215, 221, 223, 237, 238, 239, 240, 241, 242, 243, 244, 245, 246, 248, 249, 256, 257, 258, 260, 261, 262.
    See also GLAM sector
    digitisation of, 87, 238, 239, 246, 248
    remediation of, 243
curation. *See* data: curation of,
CurationProject 238

DARIAH. *See* Digital Research Infrastructure for the Arts and Humanities (DARIAH)
data. *See also* FAIR ('findable, accessible, interoperable, and reusable')
    big data 7, 135, 145, 241, 243, 244
    collection of, 33, 210, 243, 247, 248, 256, 260
    creation of, 262
    curation of, 57, 141, 236, 238, 239, 240, 241, 255, 256, 257, 258, 262
    management of, 231, 235, 236, 237, 238, 242, 243, 244, 245, 248, 249, 250, 251, 252, 255, 256, 257, 258, 259, 260, 261, 262
    ownership of, 35, 239, 240, 242, 243, 257, 261, 262
    production of, 145, 236, 237, 247, 248
    research data 6, 35, 41, 71, 74, 210, 223, 235, 236, 237, 243, 248, 249, 250, 254, 255, 258, 259, 261, 262
    reuse of, 16, 210, 223, 224, 235, 237, 239, 241, 242, 245, 252, 257, 258, 260, 261, 262
    sharing of, 118, 210, 223, 239, 241, 249, 250, 252, 253, 254, 255, 256, 257, 258, 261, 262, 263
Data Archiving and Networked Services (DANS) 74
Dataverse 256

Dávidházi, Péter 7
Day in the Life of the Digital Humanities (Day of DH) 187
Declaration on Research Assessment (DORA) 109
Department for Culture, Media, and Sport (DCMS) 83
DHCommons 34, 36
Digital Americanist 198
Digital Classicist 198, 199
digital edition. *See* medium: digital
digital humanities (DH) 1, 2, 5, 7, 10, 13, 14, 15, 18, 30, 35, 41, 59, 60, 70, 87, 89, 90, 98, 106, 111, 112, 113, 114, 118, 119, 124, 126, 127, 130, 131, 133, 137, 140, 141, 147, 153, 163, 164, 166, 168, 177, 178, 179, 185, 186, 187, 208, 209, 215, 224, 227, 250, 254, 259. *See also* digital resource; *See also* digital scholarship
    DH community 22, 23, 39, 42, 43, 117, 163
    institutionalisation of, 177, 178, 179
Digital Humanities Quarterly 117
Digital Medievalist 191, 198, 199, 200
digital object. *See* object: digital
digital publishing. *See* publishing: e-publishing
Digital Research Infrastructure for the Arts and Humanities (DARIAH) 74, 207, 209, 218, 219, 220, 221, 222, 223, 224, 225, 226, 227, 228, 230, 259, 260
    European Research Infrastructure Consortium (ERIC) 74, 209, 218, 219, 220, 221, 225, 228, 229
digital resource 63, 81, 82, 83, 84, 86, 87, 88, 90, 91, 94, 95, 96, 97, 98, 224, 253, 254, 256
    impact of, 81, 84, 86, 87, 88, 90, 91, 95, 96, 97, 98
digital scholarship 3, 4, 18, 19, 58, 61, 115, 150, 155, 163, 164, 165, 166, 167, 170, 171, 175, 177, 178, 179, 217

evaluation of, 3, 4, 61, 106, 107, 163, 164, 165, 166, 167, 170, 178, 179. *See also* metric; *See also* peer review
Digital Scholarship in the Humanities 117
Diogenes (software) 90, 91, 100
dissemination 2, 16, 26, 29, 31, 32, 33, 34, 37, 38, 41, 42, 60, 65, 70, 71, 75, 85, 90, 92, 94, 112, 113, 183, 185, 193, 195, 200, 201, 202, 208, 254. *See also* knowledge sharing; *See also* medium
doc2vec 144
Dryad 256
Duff, Wendy M. 245
Durham University 91

eBook 53, 55, 59. *See also* publishing: e-publishing
Edmond, Jennifer 246
Electronic Archiving System (EASY) 256
Elsevier 38, 67
email 186, 187, 188, 189, 197, 198, 199, 201, 202, 212, 231, 250
  junk mail 190
  spam 187
employment. *See* hiring
Engels, Tim C. E. 54
Episciences.org 71, 74
ethnography 125, 132, 150, 227
  autoethnography 131, 132, 133, 152, 154
Europeana Collections 243
European Holocaust Research Infrastructure (EHRI) 243
European Open Science Cloud (EOSC) 5, 7, 223, 224, 226
  EOSC Governance Development Forum (EGDF) 226
  EOSC-hub 226
  Social Sciences and Humanities Open Cloud (SSHOC) 226
European Research Observatory for the Humanities (EROHS) 218

European Strategic Forum for Research Infrastructure (ESFRI) 218, 229
evaluation. *See* digital scholarship: evaluation of,; *See* peer review; *See* certification; *See* metric
Excellence in Research for Australia (ERA) 184

Facebook 68, 191, 198
FAIR ('findable, accessible, interoperable, and reusable') 19, 41, 223, 235, 236, 237, 238, 239, 241, 242, 246, 247, 249, 250, 252, 255, 258, 259, 260, 261, 262. *See also* data
'FAIRification' 236, 242, 244, 247
Figshare 41, 256
findability. *See* FAIR ('findable, accessible, interoperable, and reusable')
Fischer, Frank 224
Fitzpatrick, Kathleen 43, 106, 110, 199
Flinn, Andrew 168
'forking' 65
Fortier, Paul 169
FOSTER+ 226
French Book Trade in Enlightenment Europe, The 65
Freshwater Information Management 96

Galey, Alan 125
Gans, Joshua S. 107
Garrow's Law 93
GÉANT 211
general architecture for text engineering (GATE) 91
geographic information system (GIS) 74, 96, 116
GitHub 130
GLAM sector 81, 82, 83, 86, 87, 88, 93, 95, 96, 237. *See also* cultural heritage
Global Outlook::Digital Humanities 191

Glossa: A Journal of General Linguistics.
 See Open Library of the
 Humanities: *Glossa: A Journal of
 General Linguistics*
Goldfield, Joel D. 169, 175, 179
Google
 'Google generation' 57
 Google Inbox 187
 Google Scholar 32
 Google Search 224
 Google Translate 11
Google Ancient Places (GAP) 96
Greenblatt, Stephen 49, 50, 59
Guillory, John 55

Halliday, Leah 51
Hamming, Jeanne 148
Hanlon, Ann M. 118
Harley, Diane 64, 73
Harris, Mary Dee 168
Harrison, Teresa M. 188, 199
Harris, Verne 245
HathiTrust Digital Library 96
Henderson, Margaret E. 237
Higher Education Institution (HEI) 85
High Integration of Research Monographs in the European Open Science Infrastructure (HIRMEOS) 226
hiring 50, 107, 109, 110, 113, 114, 115, 117, 168, 175. *See also* tenure; *See also* promotion
HIRMEOS. *See* High Integration of Research Monographs in the European Open Science Infrastructure (HIRMEOS)
Hockey, Susan 168
Hoekstra, Rinke 256
Horsley, Nicola 241
Hughes, Lorna M. 85, 88
Human Computer Interaction (HCI) 52
Humanist (electronic seminar/mailing list) 166, 168, 170, 171, 173, 174, 176, 190, 192, 196, 199

Humanities, Arts, Science, and Technology Alliance and Collaboratory (HASTAC) 34
Humanities Commons (HCommons) 32, 186, 202
Hyper Articles en Ligne (HAL) 71, 74
Hypotheses.org 65

impact factor (IF) 109
informational behaviour 56, 57, 252
 horizontal information seeking 56, 57
 'squirrelling' 56
infrastructure. *See* research infrastructure (RI)
In Our Time 93
Instagram 191
Institute for Research in Computer Science and Automation (INRIA) 74
intellectual property rights (IPR) 240. *See also* legal liability
International Society for the Study of Early Medieval England (ISSEME) 199
Internet. *See* World Wide Web
interoperability. *See* FAIR ('findable, accessible, interoperable, and reusable')
IPERION CH 243
Isaac Newton: The Last Magician 93

Java 138
Jisc 56, 84
Jones, Steven 123
Journal of Digital Humanities 117
Journal of Scholarly Publishing 50
 'Special Issue on Digital Publishing for the Humanities and Social Sciences' 50
JoVE 69
JSTOR 32
Jupyter Notebook 154

Kafka's Wound 92
King's College London 93, 94, 96, 116

Kirschenbaum, Matthew G. 112, 177
Kittler, Friedrich 153
Knowledge Complexity (KPLEX) 240, 241, 244, 246
knowledge iceberg 237, 246, 262, 263
knowledge production 2, 4, 11, 13, 37, 50, 52, 63, 128, 130, 131, 133, 140, 150, 153, 155, 226, 227, 236, 243, 244, 247, 248, 249, 260, 262
knowledge representation 245, 248, 260
knowledgescape. *See* knowledge space
knowledge sharing 2, 50, 58, 59, 64. *See also* dissemination
knowledge space 212, 213, 219
Knuth, Donald 151, 154

Lachmann's method 136
LAIRAH. *See* Log Analysis of Internet Resources in the Arts and Humanities (LAIRAH)
Latour, Bruno 126, 127, 131
legal liability 18, 35, 96, 112, 239, 240, 241, 242, 253
Leiden University Centre for Digital Humanities (LUCDH) 255
Let Newton Be! 93
Library of Alexandria 214
LINGUIST List 190, 197
Linked Open Data 243
listserv (concept) 4, 168, 171, 183, 185, 186, 189, 190, 192, 193, 195, 196, 197, 198, 200, 201, 202, 203. *See also* online community
LISTSERV (software) 185, 186, 187, 188, 190
Log Analysis of Internet Resources in the Arts and Humanities (LAIRAH) 4, 84, 90, 254, 255
London Review of Books 92

Macmillan, James 92
Mallet 1
Mandell, Laura 4
Manovich, Lev 123
Marino, Mark 151

Maron, Nancy 87, 99
McCarty, Willard 170, 171, 172, 190, 196
McGann, Jerome J. 119
MediaCommons Press 40
media ecology 112, 114, 115
Medieval Academy of America 199
medium. *See also* object
   digital 23, 29, 31, 33, 35, 37, 38, 39, 43, 44, 54, 70, 106, 112, 113, 165, 177. *See also* eBook; *See also* publishing: e-publishing
   print simulation 115, 117
   print 2, 9, 15, 23, 24, 25, 26, 28, 30, 31, 34, 35, 37, 39, 40, 44, 53, 54, 56, 59, 106, 112, 113, 115, 117, 126, 165, 192, 193
Medtext-l 196
Mendeley 32, 33
metric 10, 16, 38, 39, 40, 62, 66, 67, 68, 83, 85, 88, 167. *See also* peer review
   alternative metric (altmetric) 66, 68
   bookmarks 38, 39
   download counts 3, 38, 39, 90
   likes 38, 68
   link shares 38
   page views 38
   retweets 38
   Wikipedia mentions 38
   citation 3, 9, 11, 14, 18, 32, 38, 39, 67, 88, 109, 131, 184, 193, 199, 210, 254, 262
Modern Language Association (MLA) 11, 49, 61, 114, 164, 165, 175, 184
   Committee on Scholarly Editions 165
   'Guidelines for Evaluating Work in Digital Humanities and Digital Media' 114, 164
   MLA Committee on Information Technology 164
   MLA Commons 33
   MLA style 11
   Profession 61
monograph 3, 11, 17, 28, 49, 51, 52, 53, 55, 59, 63, 69, 73, 142

Morton, A. Q. 169
Mostern, Ruth 249, 252, 253

National Endowment for the
    Humanities (NEH) 106, 178
    Office of Digital Humanities 178
National Science Foundation (NSF)
    187
Nature 53, 73, 114
Naughton, John 51
Networked Infrastructure for
    Nineteenth-Century Electronic
    Scholarship (NINES) 39, 164
Network for Digital Methods in the
    Arts and Humanities (NeDiMAH)
    2, 4, 18
new media. *See* medium: digital
Newton, Sir Isaac 93
Nitti, John 168
Norfolk Medieval Stained Glass
    Project 95
North-Atlantic Population Project
    (NAPP) 256
Northumbria University 95
Notes and Queries 201
Nyhan, Julianne 168

object. *See also* medium
    digital 15, 17, 35, 37, 39, 52, 106,
        126, 128, 155, 166, 248
    epistemic 126, 130
    physical 12, 16, 52, 55, 106, 213
Old Bailey Online 88, 90, 93
Old English Newsletter 195
Olsen, Mark 174, 175
Olsen, Stephen 4
online community 183, 191, 198, 199,
    200
open access (OA). *See* publishing:
    open access (OA)
Open Access Publishing in European
    Networks (OAPEN) 53, 54, 56
OpenAIRE 226
Open Book Publishers 69
Open Journal Systems 197
Open Library of the Humanities 67

*Glossa: A Journal of General
    Linguistics* 67
OpenMethods Metablog 65
'Open Notebook History' 65
Open Researcher and Contributor ID
    (ORCID) 73
Open Scholarly Communication in
    the European Research Area for
    Social Sciences and Humanities
    (OPERAS) 226
Open Science MOOC 226
Open Science Policy Platform (OSPP)
    226
Open University 96
OPERAS. *See* Open Scholarly
    Communication in the European
    Research Area for Social Sciences
    and Humanities (OPERAS)
optical character recognition (OCR)
    248
ORCID. *See* Open Researcher and
    Contributor ID (ORCID)
Out of the Wings: The Research and
    Practice of Spanish American
    Theatre in Translation 93
ownership. *See* data: ownership of,
Oxford Multi Spectral 91
Oxford University Press 38, 184
Oxyrhynchus Online 94

Palgrave 69
    Pivot 69
PARTHENOS Hub 65
Peek, Robin 187, 188, 191, 196, 201
peer review 3, 4, 34, 39, 58, 64, 66,
    68, 71, 75, 108, 110, 113, 114,
    118, 128, 130, 155, 156, 164, 166,
    167, 170, 171, 172, 173, 174, 175,
    176, 177, 178, 179, 254, 255. *See
    also* publishing
    flaws of, 66, 108, 130
    implementation of, 172, 173
    implicit peer review 174, 175
    open peer review 40, 66, 71, 179
    organisation of, 173
Perl 138

Peters, Douglas P. 108
Pickle, Sarah 99
Planck, Max 116
PLOS ONE 113, 114
Posner, Miriam 6
Postmodern Culture 190
Priddy, Mike 240
print edition. *See* medium: print
Profession. *See* Modern Language Association (MLA): *Profession*
programming 129, 132, 134, 135, 136, 137, 138, 143, 145, 146, 149, 150, 152, 153, 154, 156, 169, 173, 229
 pair programming 142
promotion 3, 10, 50, 51, 64, 109, 110, 168, 176, 199. *See also* hiring; *See also* tenure
Public Access Computer Systems Review 190
publishing. *See also* medium; *See also* monograph; *See also* peer review
 democratisation of, 16, 67
 e-publishing 49, 53, 67, 164, 177, 202. *See also* eBook
 open access (OA) 16, 17, 29, 49, 63, 69, 70, 74, 93, 184, 186, 202
  gold 17, 70
  green 17, 70
 publishers as gatekeepers 12, 16, 17, 19
 self-publishing 16, 32, 37, 174
 traditional publishing 12, 16, 19, 33, 39, 43, 58, 69, 106, 202
Publons 68

QRator 94

Raben, Joseph 169
Ramsay, Stephen 152
Ramsey, Stephen 152
readership. *See* audience
registration 26, 27, 34, 35, 36, 37, 38, 39, 41, 42, 68
Re-imagining the Literary Essay for the Digital Age (RILEDA) 92

Research Assessment Exercise (RAE) 85
Research Excellence Framework (REF) 30, 85, 88, 89, 91, 94, 96, 97, 98, 99, 184
ResearchGate 32, 186, 202, 252
Research Information Network (RIN) 60
research infrastructure (RI) 19, 74, 207, 208, 209, 210, 211, 212, 214, 215, 216, 218, 221, 227, 228, 229, 230, 243, 259, 260. *See also* Digital Research Infrastructure for the Arts and Humanities (DARIAH); *See also* Collaborative European Digital Archive Infrastructure (CENDARI); *See also* European Holocaust Research Infrastructure (EHRI)
 cyberinfrastructure 209, 216, 217, 231
  conceptual 209
  critical 209
 definition of, 207, 209, 210, 230
 development of, 19, 74, 214, 216, 243
 function of, 212, 213, 216, 227
 necessity of, 214, 235
 symbolic value of, 208
 tactical infrastructure 209
reusability. *See* FAIR ('findable, accessible, interoperable, and reusable')
Review Journal for Digital Editions and Resources (RIDE) 39, 67
rhetoric 34, 128, 133, 146, 151, 152, 194, 229
Rockwell, Geoffrey 137, 152, 164
Romer, Paul 222
Rossetti Archive 87
Routledge 123, 184
Royal Academy of Dramatic Art (RADA) 93
Royal Shakespeare Company (RSC) 93
Ruecker, Stan 125

Russell, Bertrand 154
Sample, Mark 106
Samuels, Lisa 119
SaneBox 187
scarcity correlation 105, 112, 113, 114, 115, 118
School of Advanced Study, University of London 116
Schreibman, Susan 4, 118
Science 73, 114
Science, Technology, Engineering, and Mathematics (STEM) 36, 237
Scopus 62
Scriptura 91
Selwood, Sara 86
Shakespeare Quarterly 40
Shakespeare's Globe 93
Shepherd, George B. 107
Shirky, Clay 58
Shoemaker, Robert Brink 59
  London Lives 54, 59
Sianel Pedwar Cymru (S4C) 92
Silver Lining Theatre Company 93
Social Interpretation (SI) 94
software 1, 11, 13, 15, 16, 22, 32, 35, 36, 37, 59, 69, 90, 92, 113, 118, 123, 124, 126, 127, 128, 129, 130, 131, 133, 134, 136, 137, 138, 139, 141, 142, 146, 148, 150, 151, 152, 153, 154, 155, 169, 170, 171, 175, 185, 196, 210, 214, 224, 231. *See also* codework; *See also* coding
SourceForge 130
Standard Evaluation Protocol (SEP) 30
stemma 136, 146
Stephen, Timothy 188, 199
Sterne, Laurence 95
Strandlines 94

Tales from the Old Bailey 93
Tanner, Simon 84, 86, 88, 91, 98
Taylor & Francis 38
technology 1, 2, 3, 5, 8, 9, 14, 15, 19, 21, 29, 34, 38, 39, 70, 72, 96, 124, 127, 134, 135, 171, 176, 185, 190, 192, 195, 197, 199, 202, 211, 215, 216, 221, 227, 231, 237, 255. *See also* 'black box'
tenure 22, 50, 61, 64, 106, 107, 109, 110, 128, 148, 149, 166, 168, 169, 171, 173, 175, 176, 199, 250. *See also* hiring; *See also* promotion
Text Encoding Initiative (TEI) 39, 40, 172, 211, 214
text-mining 89, 90
The Beauty of Diagrams 93
Thesaurus Linguae Graecae 91
thick description 235, 243, 245, 246, 247, 255, 257, 258, 261, 262
Thoegersen, Jennifer L. 251
  '"Yeah, I Guess that's Data": Data Practices and Conceptions among Humanities Faculty' 251
Thomson ISI 62
THOR 69
Throsby, David 84
TIDSR. *See* Toolkit for the Impact of Digital Scholarly Resources (TIDSR)
Toolkit for the Impact of Digital Scholarly Resources (TIDSR) 84, 85
transactive memory system (TMS) 213, 222
Transcribe Bentham 94
Truschke, A. 60
tweet. *See* Twitter
Twitter 2, 4, 6, 23, 34, 59, 65, 191, 198

Ulster Poetry Project 95
Understanding Infrastructure: Dynamics, Tensions and Design 212
University of Aberdeen 96
University of Birmingham 93
University of California, Berkeley 77, 176
University of Cambridge 93
University of East Anglia 95
University of Kent 96
University of Leeds 92

University of Oxford 91, 94, 95
University of Reading 95
University of Sheffield 91, 93
University of Strathclyde 93
University of Sussex 93
University of Virginia Press 176
University of Westminster 92, 94
Unsworth, John 216, 217
Usenet 187, 188

Vectors 164
Vernon Manuscript 95
Voltaire 95

Washington Post, The 105
'water-cooler' metaphor 191, 192, 195, 196, 197, 198, 200

Web 2.0 29
WebCrawler 191
Wellcome Open Research 69
Welsch, Erwin K. 188
Whitman Archive 87
Wikipedia 10, 38
word2vec 144
World Wide Web 166, 183, 188, 197, 202. *See also* online community
World Wide Web Consortium (W3C) 211

YouTube 41

Zotero 11
Zundert, Joris J. van 132, 151, 153

This book need not end here…

Share

All our books — including the one you have just read — are free to access online so that students, researchers and members of the public who can't afford a printed edition will have access to the same ideas. This title will be accessed online by hundreds of readers each month across the globe: why not share the link so that someone you know is one of them?

This book and additional content is available at:

https://doi.org/10.11647/OBP.0192

Customise

Personalise your copy of this book or design new books using OBP and third-party material. Take chapters or whole books from our published list and make a special edition, a new anthology or an illuminating coursepack. Each customised edition will be produced as a paperback and a downloadable PDF.

Find out more at:

https://www.openbookpublishers.com/section/59/1

You may also be interested in:

Engaging Researchers with Data Management
The Cookbook
Connie Clare, Maria Cruz, Elli Papadopoulou, James Savage, Marta Teperek, Yan Wang, Iza Witkowska, and Joanne Yeomans (eds.)

https://doi.org/10.11647/OBP.0185

Text Genetics in Literary Modernism and Other Essays
Hans Walter Gabler

https://doi.org/10.11647/OBP.0120

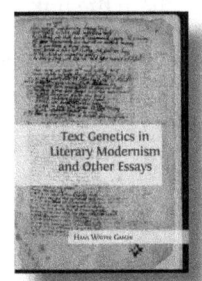

Digital Humanities Pedagogy
Practices, Principles and Politics
Brett D. Hirsch (ed.)

https://doi.org/10.11647/OBP.0024

Text and Genre in Reconstruction
Effects of Digitalization on Ideas, Behaviours, Products and Institutions
Willard McCarty (ed.)

https://doi.org/10.11647/OBP.0008

www.ingramcontent.com/pod-product-compliance
Lightning Source LLC
Chambersburg PA
CBHW051049230426
43666CB00012B/2629